MINERAL LEASING AS AN INSTRUMENT OF PUBLIC POLICY

This book is the fifth in a series based on the Economic Policy Conferences of the British Columbia Institute for Economic Policy Analysis.

Conference on Mineral Leasing as an Instrument of Public Policy
Victoria, 1974

MINERAL LEASING AS AN INSTRUMENT OF PUBLIC POLICY

Edited by

Michael Crommelin and Andrew R. Thompson

Published for
THE BRITISH COLUMBIA INSTITUTE FOR ECONOMIC POLICY ANALYSIS

UNIVERSITY OF BRITISH COLUMBIA PRESS
VANCOUVER

MINERAL LEASING AS AN INSTRUMENT
OF PUBLIC POLICY

Canadian Cataloguing in Publication Data

Main entry under title:

Mineral leasing as an instrument of public policy

(British Columbia Institute for Economic Policy Analysis series; 5)
Based on a conference held in September 1974.
Bibliography:
Includes index.
ISBN 0-7748-0058-5 bd.
ISBN 0-7748-0059-3 pa.

1. Mining leases—Congresses. 2. Mines and mineral resources—Congresses. 3. Mines and mineral resources—Taxation—Congresses.

I. Crommelin, Michael, 1945-   II. Thompson, Andrew R., 1925-   III. Series: British Columbia Institute for Economic Policy Analysis. British Columbia Institute for Economic Policy Analysis series; 5)
HD9506.A2M55   333.8   C77-002108-5

International Standard Book Number
(Hardcover edition) 0-7748-0058-5
(Paperback edition) 0-7748-0059-3

Printed in Canada

Contents

The British Columbia Institute for Economic Policy Analysis

The British Columbia Institute for Economic Policy Analysis was set up to foster independent research in public policy and to help bring the scholarly resources of the universities to bear on problems of government in specific areas: unemployment, public finance, industrial organization, and natural resource use. It acted as a halfway house between academe and action. The institute was established by the provincial government but was independent of it. It drew the income to support its functions from an endowment fund.

The institute as such did not endorse viewpoints. The only restrictions on work done under its auspices were that it bear constructively on public policy and be of high professional competence. Each individual assumed responsibility for his own findings. The institute encouraged individual researchers to develop their viewpoints into workable policy recommendations and to engage in creative dialogue with civil servants, business and labour leaders, citizen groups and public officials, and others. The institute sought to provide a forum where many doctrines might be tried.

The institute initiated and defined research topics and responded to requests for consulting. Where possible it allocated requests to researchers in the government or university system, using its contacts to serve as referral and coordinating agency.

To those ends the institute maintained a research staff, supported other scholars within the province's universities, sponsored seminars and symposia where findings could be advanced and criticized, published the results of its sponsored endeavours and offered in-service training in economic analysis to public servants. As an adjunct of the university system in British Columbia it supported students and otherwise engaged them in its activities. For the public service in British Columbia the institute engaged civil servants in its activities and encouraged a climate to attract, train, and hold professionals in government.

Walter D. Young
Chairman, Board of Directors

Acknowledgments

The discussion sessions at the conference on mineral leasing were enriched by the participation of a number of experts in addition to the contributors to this volume. In some cases, new ideas were introduced at this stage and later incorporated by authors in their revised papers. We acknowledge our debt to the following for their role in both the conference proceedings and this publication: Professor Ernst R. Berndt, University of British Columbia; Mr. R.G.S. Currie, Panarctic Oils Ltd; Dr. Anthony Hepworth, British Columbia Energy Commission; Mr. Robert Muir, practising lawyer, of Calgary; Professor T.L. Powrie, University of Alberta; Dr. Marvin Shaffer, British Columbia Energy Commission; and Professor G. Campbell Watkins, University of Calgary.

Dr. Hepworth and Dr. Shaffer also assisted the editors by contributing detailed summaries of the conference sessions.

Mason Gaffney, as executive director of the British Columbia Institute for Economic Policy Analysis, shrewdly judged that a conference on this topic would prove both challenging and rewarding. Thereafter, he was most generous in the freedom given the editors to devise the programme and invite the participants. Ms. Jean Mohart, executive assistant of the Institute, deserves special credit for her friendly and efficient handling of the considerable administrative tasks. Ros, my wife, helped greatly with preparation of the index and checking of the proofs.

Michael Crommelin

Foreword

This book is the fifth in a series based on symposia sponsored by the British Columbia Institute for Economic Policy Analysis. The first is on British Columbia forest policy, the second on pricing of local services and effects on urban spatial structure, the third on resource taxation in intergovernmental relations, and the fourth on pollution control.

The Institute for Economic Policy Analysis was chartered to underwrite research in and discussion of several aspects of policy in British Columbia. Paramount among these is the administration of Crown lands, which in British Columbia comprise 95 per cent of the land itself, plus mineral rights under much of the outstanding private land. The editors of this volume point out that many nations around the world are similarly landlords over vast mineral rights, notably in coastal seabeds.

Recent years in Canada have brought wide currency to the hitherto unfamiliar concept of economic rent, particularly in respect to subsurface minerals. Policies of the federal and the provincial governments have been and are being reviewed on the grounds that some minerals yield a surplus (that is, "rent") above costs and that this surplus is peculiarly affected with a public interest. (This economist's concept of rent applies to all resources, but in the recent Canadian context has been applied mainly to minerals.) Critics of this concept point out that minerals must be first discovered and later replaced, an issue the pros and cons of which are discussed in these pages.

In terms of policy, many issues boil down to a question of how the Crown should write leases which divide the pie among explorers, producers, and the Crown landlord. One school favours *bonus bidding*, transferring control to a lessee in one transaction, thus disposing of all frictions and disincentives caused by the Crown's later asserting residual public claims. Another school favours continuing *payments over time*, lowering front money requirements, sharing risk between Crown and lessee, and letting rent payments rise or fall, as the case may be, as time and experience reveal more about the richness of deposits. Other options are *work commitment bidding*; *property taxation*; *government involvement in exploration*; and the *relaxation of rent collection in return for its reinvestment*. All these are discussed in this volume with vigour and style by their advocates.

Editors Andrew R. Thompson and Michael Crommelin are both university law teachers—one Canadian and one Australian. Their nationality and profession have not prevented their bringing together here an international and interdisciplinary group composed primarily of economists, but including mining engineers, businessmen, and industry consultants. The result is a mix of applied and institutional economics to delight the modern followers of John R. Commons, Thorstein Veblen, and Clarence Ayres, as well as the "Law and Economics" scholars, whose numbers are currently burgeoning in several universities. It is an outstanding contribution to a rapidly growing field of study.

Mason Gaffney

Introduction

MICHAEL CROMMELIN
ANDREW R. THOMPSON

Events in recent years have once again raised the spectre of shortages of various commodities. At the forefront of this concern has been energy resources, principally oil and gas, and other minerals. In many cases, threatened or actual shortages have led to sharp price increases. All this has resulted in a renewed emphasis upon the management of existing mineral resources.

British Columbia is a major producer of a number of minerals. The mining industry in the province lays claim to the title of "B.C.'s second industry," conceding preeminence only to forestry. Within the industry the greatest revenue is derived from metal mining, particularly of copper, but the province also has a substantial endowment of coal, natural gas, and (to a lesser extent) petroleum reserves.

A change in the provincial government in 1972, together with the subsequent "energy crisis" and world mineral price increases, paved the way for a series of amendments to British Columbia mining and petroleum laws. The feature that these changes had in common was an increased burden upon the mining industry. Opposition, by no means confined to the mining industry, was swift and vocal. As a result, the question of management of mineral resources was brought into a sharper focus than at any time during the previous history of the province. Of course, in this regard British Columbia merely exemplified the worldwide reexamination of basic attitudes towards natural resources.

In September 1974, the British Columbia Institute for Economic Policy Analysis sponsored a conference on mineral leasing, in line with its mandate to conduct independent research in the public service. The conference took the form of a meeting of experts drawn from universities, governments and industry, at which papers were presented and subjected to searching criticism. This present volume is a collection of articles based on these papers. Each article is the result of considerable review in the light of discussion that took place at the conference. In those cases where oral comments added a further dimension to an article, either by highlighting a significant difference of opinion among conference participants or by raising an issue not dealt with by the article itself, commentators were prevailed upon by the editors to commit their remarks to writing for inclusion with the relevant article. In addition, this volume contains an edited version of the "open" session conducted at the conclusion of the formal proceedings, together with a final note by one of the editors on the most contentious subject of the conference, economic rent.

Before reviewing the articles, however, it is desirable to clarify several matters regarding the scope of the volume and its organization. First of all, the articles and proceedings are not relevant only to British Columbia. The preferred approach has been to devote the majority of the book to a consideration of general problems encountered in mineral leasing and, to this end, to call upon the combined wisdom of experts drawn from many jurisdictions. In the concluding sections these problems and their possible solutions have been placed in the context of the specific conditions prevailing in British Columbia.

Secondly, an important limitation was placed upon the deliberations. The primary objective of the conference, as suggested in its title, was to consider the problems of mineral leasing from the point of view of a government in the position of owner of mineral resources in the ground. Attention was directed to the difficulties faced by such a government in its dual capacity of owner and regulator. This conforms with the position in British Columbia where the vast preponderance of undeveloped mineral resources are vested in the Crown in right of the province. However, this is not unique to British Columbia; many of the other Canadian provinces are in a similar situation, as are the Australian states and the federal government of the United States, so far as the public domain lands are concerned. Moreover, the coastal nations of the world all find themselves in this position regarding mineral resources of the seabed within national jurisdiction. The implications of this dual capacity have not been fully explored, but one factor is clear: the obligations of ownership give rise to a concern for the distribution of the benefits generated by resources development which may not be inherent in mere regulation.

Thirdly, the scope of the volume should not be viewed too narrowly. "Mineral leasing" may encompass a range of subjects, extending from the tenure arrangements under which private rights are acquired over public resources, through specific taxation provisions affecting the mining industry, to the general financial climate in which mining rights are exercised. All of these come within the topic for discussion. Questions of public policy seldom allow an arbitrary line to be drawn separating specific tenure arrangements from more general resource management issues, and it was felt that little benefit was to be gained from such an attempt.

Fourthly, the articles in this volume generally follow the same author-presentation sequence as the prearranged programme at the conference. The first part is devoted to a discussion of alternative leasing procedures, both competitive and noncompetitive. The shortcomings in these mechanisms receive further and more detailed attention in the second part, where specific problems inherent in devising a mineral leasing policy are examined: the timing of resource development, the conduct of exploration, the role of public enterprise in the management of mineral resources, intergovernmental competition for mineral resource revenues, and selecting the appropriate taxation regime. In the third part, the mineral leasing policies currently in force in British Columbia are outlined and evaluated in terms of the criteria suggested in the preceding articles. The final section contains the edited transcript of wide-ranging and informal discussion of issues still unresolved.

Thirteen articles make up this volume. The main arguments developed in these articles are summarized hereunder.

Mason Gaffney's provocative and stimulating article emphasizes the importance of institutions in motivating individuals to enhance the public interest. He says that the objective to be sought by a government, in its capacity of mineral resource owner, is the maximization of the present value of the economic rent derived from minerals. He identifies eight mistakes that a government may make in attempting to achieve this objective: overdecentralization; overdelegation; the allowing of excessive premia for risk; the allowing of open access to minerals by explorers; the underpricing of primary products; the confusion of rent with profit; the overlooking of nonmining taxes; and the overconsolidation of accounts. The article is devoted to an analysis of these mistakes with particular reference to their effects upon the quantum and distribution of mineral rents. Several positive exhortations emerge from the discussion. These include making the present value of rent, rather than income, the base for levying provincial mining taxes; keeping capital requirements for operating as low as possible; making private tenures short to prevent tying up of reserves; taxing reserves themselves through a delay rental proportional to their value; and controlling the timing of resource development to maximize the present value of rent.

Frederick Peterson's first article, "An Economic Theory of Mineral Leasing," develops a theory of optimal exploitation and leasing under ideal conditions. In this model, any number of policies or combinations of policies can be used by the government to capture the present value of its mineral properties. Leasing policy is not important. Tracts can be offered for lease or sale individually or all at once and at any time, so long as resource extraction is not delayed and the timing of exploration and exploitation is not distorted. The market produces the time path of extraction which maximizes present value for each tract. However, the "ideal" conditions which are required for the development of this model are extremely restrictive. Recognizing this, Peterson goes on to consider policy implications without the earlier assumptions regarding availability of information, a large number of bidders, no monopoly effect on mineral prices and no alternate uses of mineral properties. In these conditions, the market no longer assures the optimal timing of resource extraction for the maximization of the present value of government revenues. Peterson concludes that mineral leasing policy must operate in a complicated economic and political environment; no policy can be rejected out of hand because it fails the optimality tests of the ideal model.

In "Cash Bonus Bidding for Mineral Resources," Walter Mead identifies three problems to be solved by any mineral leasing system. First, the system must determine who is to be given the right to exploit publicly owned resources. Second, a price must be fixed for mineral rights transferred by the government to a private operator. Third, the leasing system should encourage an efficient method of production of minerals. Mead prefers a competitive leasing system to the alternative of government negotiation with private operators. Moreover, where competition exists among private operators for the acquisition of mineral rights from a government, Mead advocates a cash bonus bidding system, such as that used by the United States government upon the Outer Continental Shelf lands for oil and gas leasing, rather than the alternative of royalty bidding. He also discusses the factors that are important in choosing between oral and sealed bidding methods.

Gregg Erickson's article on "Work Commitment Bidding" examines the method of mineral leasing whereby exploitation rights are acquired by the operator prepared to commit himself to spending the greatest sum in developing the resource. The basis for comparison is the system advocated by Mead, that of cash bonus bidding. Erickson recognizes that a critical factor in this comparison is the objective chosen by the government. If the success of a mineral resource management policy is measured by the physical quantity of the mineral educed from the earth, Erickson concludes that the work commitment bidding is clearly superior to cash bonus bidding. However, if success is measured in terms of the contribution made by a publicly owned resource to economic welfare, work commitment bidding involves a subsidy to mineral production. In this case, any argument for the adoption of such a system must first establish that a subsidy is required. In addition, it must show that the work commitment approach is the least costly method of providing the required subsidy. In this respect, Erickson's conclusion is interesting. He illustrates that the amount an operator will promise to spend under the work commitment system will *exceed* the sum of the bonus and the development costs that the operator would have been prepared to spend on the tract under a cash bonus bidding system.

Ross Garnaut and Anthony Clunies Ross in "A New Tax For Natural Resource Projects" propose a particular form of tax that is progressive with respect to the rate of return on investment as calculated by discounted cash flow methods. This tax, called the "Resource Rent Tax"(RRT), is designed to ensure that the private operator obtains a specified after-tax rate of return on total cash flow before the government shares in mineral resource revenues. In this way, it reduces the amount of risk borne by the private operator engaged in a mining venture, and correspondingly increases the government's share of risk. Furthermore, in specifying the private operator's return prior to the development of mineral resources, the RRT creates less uncertainty than revenue collection systems which are dependent upon an *ex post* determination of the size of the economic rent. The authors recognize the very important point that the size of the economic rent is not independent of the mechanism employed by the government for its collection. They argue that the RRT, by shifting risk from the private operator to the government (the private operator is assumed to be risk averse while the government is not) and by dispelling uncertainty, allows the economic rent available for collection by the government to be larger than under alternative leasing arrangements. Finally, Garnaut and Ross point out that the RRT is applicable to a situation where competition among private operators is insufficient to allow use of bidding systems, although they do illustrate how the tax may be modified for use in a competitive environment.

Brian Mackenzie in "Investment in Information for the Assessment of Mineral Value: Some Guidelines For Mineral Leasing Policy" also recognizes the relationship between the government revenue collection mechanism and the size of economic rent generated by mineral development. His article describes the stages, information responses, and decision criteria associated with private investment in mineral exploration and development. He thus identifies three issues which are of critical importance to the private operator. These are (a) the average exploration cost associated with making a discovery; (b) the probability of discovering an economic mineral deposit; and (c) the return resulting from an economic discovery. Mackenzie advocates the establishment of a data base in any mineral region to allow empirical analysis of these issues. Finally, he considers the effect of different government revenue collection systems upon the private operator's view of these three issues and advocates reliance upon profit taxation on the ground that it interferes least with the private investment process.

In his second article, "The Government Role in Mineral Exploration," Frederick Peterson argues that the government should perform or contract out the early exploratory work on public lands because three problems prevent the private market from functioning properly. The first of these, described as "information spillovers," results from the lack of correspondence between mineral tracts and subsurface geological conditions. Exploration on any one tract thereby yields information of value in other tracts. In the absence of agreements among owners of adjoining mineral rights, there will be a divergence between the social and the private benefits derived from private mineral exploration. The second problem is one of economies of scale. Peterson suggests that mineral exploration is a classic decreasing-cost industry; thus, a government should generate and disseminate exploratory information for a nominal fee in order to utilize scale economies. The third difficulty concerns the presence of risk in exploration. Asserting that private operators are averse to risk, Peterson argues that government involvement is necessary to produce the optimal level of exploration and desirable to increase the level of competition among private operators in the later stages of mineral resource development. Peterson also considers arguments against government involvement in exploration, but concludes that these arguments are not of sufficient strength to overcome the factors calling for such involvement.

On the other hand, Arlon Tussing in "The Role of Public Enterprise" concludes that "governmental ownership of producing operations is not generally the most effective way of accomplishing the social ends for which it is currently being advocated in these industries." While he does support the case for government ownership of undeveloped land and natural resource stock, he is less sanguine about the prospects of state enterprise in the business of developing and producing minerals. He lists—and discounts—three economic arguments that are commonly put forward for government ownership at these stages of development. The first of these is that the state should establish or maintain productive activity that would not be profitable as private enterprise but whose external benefits are deemed to justify a subsidy out of the public purse. Tussing feels that the mineral industries of Western Canada have scant claim for public subsidy, and, of course, a subsidy does not require state ownership. The second is the perceived inability of private enterprise, because of the great size or risk of the venture in question, to assemble sufficient capital. Again Tussing feels that this has little application to British Columbia. The third argument is that government ownership of exploration and production facilities is one means of preventing the exercise of natural monopoly power in the mineral industries, or, alternatively, it allows the public collection of monopoly profits or resource rents that would otherwise be captured by private enterprise. On this point Tussing agrees that direct government involvement is one way of achieving the desired objective; but he suggests that it is likely not to be as effective as the combination of a leasing system that takes full advantage of competition among private firms with an appropriate tax system. Nevertheless, if public enterprise is to be established in the minerals industry, Tussing feels that there are a number of rules that ought to be followed to avoid shortcomings in its operations. The purpose of the enterprise should be clearly defined; the enterprise should not operate as a monopoly; it should not be given sovereign immunities; both the general public and staff of the enterprise should be given a material interest in its success and efficiency; the policies of the enterprise should be responsive to public policy without the threat of day-to-day political intervention; the enterprise should be under pressure to pay dividends; a distinction should be maintained between the enterprise and the government in its capacity as a landowner; and the enterprise should take advantage of services offered by private operators in the mineral industry at competitive prices.

David Quirin and Basil Kalymon identify three distinct issues within the subject of their article, "The Problem of Timing in Resource Development." The first, described as the "scheduling problem," concerns the selection of the best producing rate in respect of an investment made at a given point in time. In view of earlier treatments of this subject, the authors do not pursue it in their article. The second, called the "problem of historical timing," deals with the selection of an appropriate time for undertaking a given project considered in isolation. The third, the "problem of cyclical timing," concerns the selection of investment priorities between competing projects which might, on grounds of historical timing alone, all be undertaken simultaneously, but which cannot because of bottleneck constraints. It is significant to note that Quirin and Kalymon choose as their objective the maximization of social welfare rather than government revenue from mineral development. With regard to the question of historical timing, they argue that, contrary to the tenor of Peterson's conclusions in "An Economic Theory of Mineral Leasing," there is very little evidence to suggest that intervention by public authorities will improve the timing of resource development decisions over that which would result if market forces were allowed to operate without the premature development bias implicit in present tenure regulations. So far as the cyclical timing problem is concerned, though, they feel that there is at least a *prima facie* case for the exercise of some administrative discretion in selecting investment priorities, and they suggest guidelines that should be employed in exercising this discretion.

In describing John Helliwell's article, entitled "Overlapping Federal and Provincial Claims on Mineral Revenue," it seems best to repeat the questions that he asks as the outset. These are: How does divided jurisdiction affect the nature and efficiency of natural resource management? What are the effects of competition between jurisdictions? Is there any "right" division of authority and resource revenue between federal and provincial governments?

In setting out to answer these questions, Helliwell looks specifically at the Canadian federal system. He describes several types of overlapping claims on natural resources and then suggests how these overlaps affect the nature and operation of various types of mineral leasing policy. For example, he takes four general categories of leasing policies (gross royalties, net royalties after deducting actual costs, royalties or payments based on estimated costs and revenues, and taxes or regulations governing resource use) and investigates how each of these interact with a federal corporation income tax. The actual provisions of the Canadian federal budget of 6 May 1974, which raised a storm of protest in the mineral producing provinces, are used in this analysis. He then goes on to investigate the effect of the federal export tax on oil. Next, he considers the lengths to which interjurisdictional conflict could proceed and, finally, reviews revenue sharing and equalization payment possibilities that could be used in a more cooperative environment. His views on this subject, while strictly applicable only to the particular Canadian situation, are nevertheless of great interest. He suggests that the federal government should back right out of any special tax treatment for the resource industries, leaving the management function in the hands of the provinces. Two-price systems should be subject to close scrutiny. Limitations should be agreed among jurisdictions upon subsidies for local use of resources. The anomalous role of resource revenues in the Canadian equalization payment system should be altered so as to include all resource revenues in the base used to calculate payments into the equalization fund as well as entitlement for equalization payments.

The theme of Paul Bradley's article "Some Issues in Mineral Leasing and Taxation Policy: Preface to a Simulation Study" is that the selection of a particular set of mineral leasing and taxation measures should be made with reference to long-term government revenue maximization. This, he says, is to be contrasted with capture of short-term economic rent. However, when the goal of long-term revenue maximization is chosen, the appropriate mineral leasing and taxation policy is far from obvious. The presence of three types of risk—commercial, geological, and political—takes the situation a long way from the "perfect world" in which economic theories are so frequently developed. The result is that a choice cannot be made between alternative leasing and taxation mechanisms, such as bonus bidding and payment of gross royalties, purely upon qualitative grounds. Research is required into a number of questions before public policy can be formed with confidence. Bradley lists these questions, and describes a simulation programme on which he is working in an attempt to obtain long-term answers.

In "Mineral Leasing in a Private Enterprise System" J.L. McPherson and O.E. Owens describe the tenure and taxation arrangements by which private operators acquire exploration and production rights over metallic minerals in British Columbia. The tenure provisions have their origin in the gold rush days of the last century. A "free miner" may enter, locate, and prospect upon private and crown lands for all minerals other than oil, gas, and coal. In recent years, though, substantial changes have been introduced into this system. What previously amounted to an automatic right to mine has been replaced by ministerial discretion, and gross royalties have been added to the previously existing mineral income tax. The authors describe these changes in detail. They then discuss a variety of issues of importance in designing a mineral leasing policy. These include the nature of reserves; differences in exploration between metallic minerals and oil and gas; the relevance of the term "economic rent" to mineral leasing; and risk in the mining industry. They conclude with a critique of the present mining legislation in British Columbia and suggest that the public interest in mining is best served through imposition of a single tax on earned income.

The subject of Dale Jordan's article is clear from its title, "Petroleum Leasing in British Columbia." The author gives a detailed account of the system of granting oil and gas rights on crown land in the province and makes useful comparisons with the rights available in Alberta and Saskatchewan. The prospect of interjurisdictional competition for scarce exploration and development funds is again raised. The author argues that the most important factor motivating the small-scale explorer and producer, now responsible for the bulk of operations in Western Canada, is the cash flow from discoveries, and he suggests that leasing and taxation arrangements be designed to take account of this fact.

During the final "open" session at the conference itself, it was widely acknowledged that three issues had dominated the proceedings but had remained largely unresolved: the concept of economic rent; the importance of risk and uncertainty and how they should be dealt with by governments; and the scope for direct government intervention in mining at both the exploration and the production stages. The record of this session (the fourth section of this volume) confirms that consensus was never attained and aptly illustrates the range of views which must confront any government during the formulation of its mineral leasing policies.

The concluding note on "Economic Rent and Government Objectives" was prepared by Michael Crommelin after the conference and sets out to provide a bench mark against which the conflicting views of economic rent (so apparent in the articles and discussions) can be evaluated. It traces the historical development of this concept from Adam Smith to the present day, notes inconsistencies as they occurred from time to time, and adopts the generalized approach of Joan Robinson. The next step is to apply this concept to the mining and petroleum industries. In so doing, Crommelin endeavours to clear away (or at least clarify) many of the points of difference which emerged so starkly at the conference.

PART ONE

Alternative Leasing Procedures

Objectives of Government Policy in Leasing Mineral Lands

MASON GAFFNEY

The statesman who should attempt to direct private people in what manner they ought to employ their capitals, would not only load himself with a most unnecessary attention, but assume an authority which could safely be entrusted to no council or senate whatever, and which would nowhere be so dangerous as in the hands of a man who had folly and presumption enough to fancy himself fit to exercise it.

Adam Smith

The statesman in charge of leasing vast crown lands might seem forced to direct private people how to employ their capitals, and their persons, too. He certainly has the power, and he certainly has the responsibility to use his power in the Crown's interest, but there is a way to do so without being arbitrary, capricious, meddlesome, subjective, tyrannical, or inefficient. To serve his citizens best, the statesman should act much like a private landowner maximizing his net income from lands. He should resist the temptation to use his power to manipulate and control, foster and suppress, divert and channel, reward and punish on the too easy presumption that the market has no rationale or normative value of its own. Generations of economists have established that it has, and governments seeking to improve on it need face a certain burden of proof. A landowner maximizing the net income from lands is tolerably likely, thereby, to be directing them to their highest and best use—that of meeting the most human wants and needs. Net income, after all, is a measure of the excess of benefits over costs.

The official who grasps that concept may then identify many costs that some people dump on others and benefits they bestow on each other. He may seek to internalize these externalities in his planning. But as one surveys the dogmas that hold sway in many professions concerned with land use, one sees a dozen bad ones for every good one. It is the rare official today who can sort these dogmas out well enough to improve on the market. This article is an attempt to help with the sorting. But the improvements that are possible consist mainly in helping the market work better, not in rejecting it.

In Canada a province that owns land may be subject to federal sanctions when it sets about collecting resource revenues. In 1974 John Turner, the minister of finance, advanced a budget in which provincial royalties were made nondeductible for federal income tax. The government fell, but the ensuing election made the new rule stick. A province must proceed with one eye on the constant power struggle on this front. We will show, however, that the most economic ways of collecting rent are also the least vulnerable to being declared nondeductible.

A maximizing landowner does not simply "maximize rent." First, there is intertemporal interdependence of rent, and the objective is to maximize the present value of net land income over time, not the rent of *any* one year. With minerals especially, timing is of the essence—not just when to produce but also when to explore, when to begin, how fast to produce, and when to stop. Second, there must be a rent base, which presupposes public investment in infrastructure and private and/or public investment in exploration. From the provincial viewpoint this need is eased by the over-proclivity of Ottawa to subsidize both exploration and infrastructure, and the province will frame its best strategy vis-à-vis Ottawa by knowing how to exploit this knee-jerk in the other team. But, whoever does it, it must be done.

Third, we must find ways to collect rent while preserving its total amount. This means avoiding heavy dependence on high gross royalty rates which destroy marginal incentives. We need other means to collect most of the rent, for rent is 60 per cent of the gross value of very low-cost mines, and 0 per cent of marginal ones.

Fourth, the land administrator should avoid dissipating rent by fostering or allowing excessive and premature investment. Land can be overdeveloped and prematurely developed as well as the reverse, and it is net, not gross rent one should maximize. Our "free mining" regulations make prospecting something like fishing on the open seas with unlimited entry, attracting new boats—or prospectors—until the average entrant can earn nothing above real costs, so there is no rent remaining. At the same time the land manager who puts reserves in cold storage should distinguish clearly between merely establishing tenure control to maximize net rent and the wielding of market power to raise prices. The latter approach is of negative social benefit when consumers are included and is of questionable benefit to the would-be monopolist unless he be exceptionally lucky and astute.

Fifth, we should avoid dissipating rent by letting lessees keep it on condition they plough it back into mining or exploring. This is in effect treating a capital investment in Mine B like a current expense of Mine A. It diverts rent from lessor to lessee and force-feeds capital into new mines below the social opportunity cost. Again, we should avoid dissipating rent by allowing rebates for domestic refining and consumption. The purported goal of subsidizing domestic secondary processing is to generate employment. But cheap feedstock induces substituting feedstock for labour. To foster jobs we should allow a wage credit, not a raw material credit. Fostering secondary industry to "develop resources" is a way of dissipating rent, not maximizing it.

Sixth, we should check and control the common itch just to meddle and manipulate. It is said that excess profits are either competed away or "imputed away" as rents. To that we should add they are often piddled away. A common kind of frittering arises from indulging the uneconomic ideologies of officials who control resources. A classic example is the "even-flow" doctrine of the forest service, embedded in the Hanzlik Formula.[1] Mining laws are replete with provisos where you pay less in return for doing this or that with supposed benefits to society. Redistributing wealth is a frequent rationale. Many a man with a little market power rationalizes his policies by assuming the mission to play Robin Hood, or perhaps God's avenging angel, within his little sphere, rather than use his resources efficiently. The result is neither efficiency nor equity, because equity far outreaches any one local industry. The official usually has no commission to impose his subjective concepts of equity on others and may only be putting a good face on self-interest in any event. The manager serves society better simply by maximizing the bottom line.

Subject to such provisos and understandings, the objective of government policy is to maximize rent and then, of course, to collect it. Rent is by definition a surplus above the return required to motivate production. It is equally well defined as the return imputed to land. In either concept it is essentially the fat without the lean. The less of the lean one cuts into by clumsiness, the more of the fat he can secure without impairing functional incentives.

The positive art of securing rent from minerals is the subject of other articles in this volume. Here I begin by clearing the ground of common and characteristic errors and blunders, errors embedded deep in our institutions, rhetoric, and cultural baggage; errors that preclude any rational effort to maximize welfare. I define eight of them: (a) overdecentralization, a hornet's nest of at least ten blunders likely to be committed in the effort to collect rent; (b) overdelegation of public authority to private giants; (c) overallowance for alleged risk; (d) overadmission of prospectors; (e) underpricing to domestic users and consumers; (f) confusion of rent and profit; (g) overlooking the taxation of nonmining activity; (h) overconsolidation of accounts, letting the strong hide behind the weak as to equity, and the weak behind the strong as to viability.

OVERDECENTRALIZATION

Overdecentralization is a transcendent bias in resource institutions. The bias makes us produce too little too late from rentable, superior deposits, too much too soon from marginal and unripe deposits and, on the whole, carry excess inventories of half-developed deposits and fixed capital along with inadequate working capital and flexible capacity to meet surprises. Overdecentralization has several causes and aspects.

Regional Development

The positive value of pushing back frontiers and opening new land is a notion engraved deeply on the cultural subconscious, inherited from generations of aggressive, landgrabbing ancestors. Having already grabbed more than our share, it is time in today's crowded world that we recognize this as overdone and concentrate on utilizing what we already have.

What is it we are trying to accomplish by regional development? Development and growth are not ends in themselves. The tundra lies peaceably until we need it; there is no call to occupy it just "because it is there." As human settlement expands we reach a margin where there are negative returns from more land development. To reach beyond that point is folly, and no less so because our ancestors did it.

Are mines the first wave of civilization? They are outriders all right, but they lack the power which agriculture used to have (when it was more labour-intensive) to pull much behind them. Mines are isolated and narcissistic, not seminal. Mine location is determined by geology with little relation to other resources and where people like to live. Output moves long distances to market. It is a truism of geography that mines create no great cities.

Mines do require much ancillary capital: rails, roads, power supply, ports and superports, unit trains, and pipelines are examples. But these are not shared much with other industries; they tend to be single-purpose.

In the philosophy of regional development private investment exerts great leverage over public investment. When someone finds a deposit way out in the bush the public chips in by extending utility and transportation networks. The marginal extensions are usually heavily cross-subsidized by the consolidated account bookkeeping that characterizes such networks. If regional development remains the lodestar, there can be no end to the submarginal extensions until they have completely drained off the surpluses generated by the rich territory of the systems. Thus, a good deal of the new capital that mines do pull behind them is a hidden subsidy of negative social benefit.

Mines do, of course, attract smelters, refiners, and reducers. But separating metal from oxygen consumes more energy per dollar of value added than any other industrial operation. Aluminum consumes some 133 kwh per dollar of value added, compared to about 10 in industry generally. Thus processing ores ties up great blocks of scarce energy and witholds it from higher uses. When the smelters are running they are known for their high load factors. This achieves some saving of capital, usually cited as a plus for the smelters. But in terms of regional development such savings accentuate their narcissism. With one consumer using the whole capacity of a power source or line there is no need to diversify the market to share the capacity. So again, subsidizing isolated mines can hardly be justified on the grounds of spillover benefits fostering regional development.

Smelters and refiners also consume another resource whose value has suddenly rocketed from the neglected to the exaggerated: the environment. A noisome smelter destroys the amenity value of real estate for miles around, which is to say the smelter owners have, in effect, appropriated part of the resources owned by others. Blighting the land repels people and discourages regional development.

As to employment, mining, often described as British Columbia's second largest industry, employs only 1.2 per cent of the labour force. Smelting and refining, too, are capital-intensive operations with a low share of labour cost per dollar of value added. The property income goes to develop those other regions like Palm Springs, Palm Beach, and Point Grey, where investors live.

Discriminating against On-stream Operations.

Governments that control price often pay less for "old" than "new" production. Prices may be controlled by a regulatory commission (like the United States Federal Power Commission) or by a government monopsony (like the British Columbia Petroleum Corporation). The British Columbia Petroleum Corporation pays 20¢ per mcf for old and 35¢ for new gas, (raised to 35¢ and 55¢ late in 1975). Where royalties are used, we sometimes find lower rates applied to newer wells or mines. Saskatchewan uses the tax mechanism so that new oil brings higher revenues to the producer than old oil. There is a tough surtax on prices over $3.38 a barrel, with exemption for new oil.

James R. Nelson labels this genus of policies "chronological marginalism."[2] It is a splendid example of the fallacy of identification. Old gas is identified with low-cost, rent yielding gas, and new gas the reverse. There is a half-truth in it when (real) gas prices are rising, because newer gas may have required the higher price to meet higher costs. But half-truths are not good enough. For collecting rent, chronological marginalism is like performing surgery with a rusty tin can.

The high royalty (or low price) in the case of older gas (or oil or ore) causes "high grading" there—marginal gas or ores are left in the ground. Even if old gas is really lower cost, it is so only on an average. Low-cost deposits have high-cost margins, and the policy in question here shuts in these intensive margins of the old deposits in favour of diverting effort to

new ones. The results are overdecentralization, needless scatter, and
territorial expansion.

An important aspect of high grading is "slow grading." That is,
production plans are shifted to the future because when the fisc takes a
gross percentage off the top from a mine, the percentage growth rate of the
owner's equity rises by a leverage effect, if (as is normal) rising demand and
falling real costs lie ahead.

D. Gale Johnson has pointed out that a share tenant has every incentive
to take as much land as his landlord will allow, and skim the cream.[3] There is
no cost to the share cropper if he uses more land; therefore, he substitutes
land for effort in every trade off. He high-grades the land and spreads out.

The Indonesian system proposed for the United States by Senator
Bentsen is a variation of chronological marginalism. Here, royalties are
low until the lessee recovers his exploration and some other capital costs
(which may mean a range of things); then royalties slide up to very high
rates. There is a rough intuitive appeal—another half-truth—in the idea;
however, it is only a variation on the rusty tin can analogy. Production
from old units is choked off in favour of getting new ones off to a roaring
start. The result resembles the lives of so many people: too many starts, too
few completions.

Yet another variation is the system currently being introduced in
Manitoba, which operates a bit like utility regulation. A firm is given a
"profit base" equal to a percentage of its unrecovered capital. The tax rate
on base profits is low, followed by a steep surcharge on excess profits. As
capital depreciates the profit base falls. Utility economists will recognize the
familiar setup that opens the door to "Averch-Johnson Effect."[4] The profit
base is like a utility rate base. To maintain the base, the firms will sink capital
into new ventures including many that are subeconomic and/or premature.
Instead of collecting rent, Manitoba will subsidize waste as it forces the
overdecentralization of mining.

Overallowance for Amortization

The private owner of minerals is likely to programme his extraction rate
so as to maximize the present value, as he should. The lessee, however, is
only too happy to tie up crown owned reserves on the cheap, if the Crown
allows. Then the speculative gains accrue to the lessee while the holding
costs, invisible to most citizens, are borne by them.

A frequent route to this end is to overstate required amortization periods.
"Long production runs" are worth a lot when "long" means say five years
instead of two. But here we are talking about decades, and it makes a big
difference, because when decades are involved the cost of interest totally
dominates the cost of capital regardless of the length of the production run.

A passing acquaintance with tables of compound interest is enough to make the point. The cash flow required to amortize a capital outlay in twenty years is only a little more than the flow needed to pay interest on it in perpetuity. That is, if you could let a lessee tie up infinite reserves, so his capital cost could be spread over all future time, his annual cost of capital would not be much lower than if he had to recover it all in twenty years. For example, if the interest rate is 10 per cent, then an investor need get only 11.75 per cent yearly for twenty years to recover his principal plus 10 per cent on the unrecovered balance.

Continuing the example, if we give him double the reserves, so he can spread his cost over forty years, we only lower his yearly cash flow requirement from 11.75 per cent to 10.23 per cent. This gain is negligible relative to many other factors involved, such as claimed risk premia which run up allegedly "required" or "target" rates of return by many percentage points. Table 1 provides some more data on amortization.

TABLE 1

CASH FLOW REQUIRED TO AMORTIZE $1 IN <u>N</u> YEARS.*

(1)	(2)	(3)	(4)
n	c.f.	Decline of c.f.	Decline of c.f. as percentage
10	0.163	—	—
20	0.117	0.046	28
30	0.106	0.011	9
40	0.102	0.004	4
∞	0.100	0.002	2

The tabulated data were derived from the following equation:

$$c.f. = \frac{i}{1 - e^{-ni}}$$

where, c.f. = cash flow
i = 0.10
e = base of natural logarithms
n = number of years of amortization period

*Also known as the "capital recovery factor," "the annuity whose present value is one," and "the instalment plan."

In fact, in twenty years a miner is likely to replace much of his capital anyway, piece by piece. What you accomplish by giving him forty years' reserves is to delay half your production by twenty years, and half your crown revenues. On the plus side, you may defer some of his costs for twenty years (somewhat less than half of them) but there remains a net loss, the delay of his revenues above costs and the delay of public revenues.

The world of public leasing is rife with devices to lock up rentable reserves. "MER"[5] regulations are notable. They force the better deposits to be produced slowly, thus forcing premature recourse to remote and marginal deposits. In locking up rentable reserves one loses not only the time value of the reserves themselves, but also that of the capital used to discover, prove out and extract them. That is, the early overhead capital has to be recovered slowly. Interest accruing on this capital cuts deeply into the net rent available for the crown landowner to tap, by whatever means.

Well spacing regulations are of like effect. They require that each unit of capital combine with more reserves, over more years. Wide spacing on rentable reserves reduces output per year, inducing premature recourse to other deposits. Of course *some* spacing requirement or unitization is needed for a common pool, but it is a question of how much.

Another bad practice is the one Alberta followed during 1950-64 of allocating production allowables in proportion to well capacity and well depth.[6] This is the way many cartels operate, and, of course, the practice encourages the construction of otherwise useless capacity to claim quotas.

Another device, whose ostensible purpose is unclear, is allowing vast area units on permits and letting a permittee or licensee or lessee hold the whole area by token activity or by production from a small part. This pattern has marked coal leases in British Columbia.

In addition to wasting rent by wasting capital, policies that slow down capital recovery sequester extra capital in a form minimally complementary to labour. Basically, capital combines with labour and makes jobs when it turns over, that is when capital is recovered and reinvested. Policies like prorating to MER force each unit of capital to lie passively a long time before being recovered. New wells, instead of being financed with capital recovered from old ones, now have to tap outside capital, drawing more and more into the industry. Slow capital recovery to investors is the counterpart of slow delivery to consumers. The latter maintains prices, again forcing recourse to marginal deposits, accentuating the industry's drain on capital from outside. All this helps to increase national unemployment rates by reducing the flow of gross investment, and hence payrolls, associated with the finite fund of national capital. In short, witholding rentable reserves wastes them by delay of use, wastes capital by misallocation, and wastes labour by reduction of job demand.

Federal Tax Provisions

The federal income tax is heavy on what are now called "tax expenditures," that is, the income tax gives preferential tax treatment that subsidizes exploration. Expensing of exploration investments is a major subsidy. Most people fail to appreciate the extent to which time is of the essence in tax matters. The privilege of deducting capital outlays currently is worth so much that it amounts to 100 per cent tax exemption. The reasoning is all simplicity. The investor lowers his present tax liability, and hence his investment, by an amount equal to the tax-rate percentage of his outlay. Later he shares the cash flow with the government on the same percentage basis. So on his reduced investment he recovers principal and interest free of all tax. The fisc only earns a return on its own investment.

A depletion allowance is offered to mineral and petroleum producers, but to obtain it in Canada they must invest in new wells or mines. This requirement of new investment makes depletion allowance less stimulating to extant producers by limiting their control over the funds retained. It arrogates part of the stimulus to the purpose of making money cheap for explorers. Thus a large flow of annual capital from producing wells is simply forced into exploring and developing new areas and deposits, even though much of it might have higher alternative uses. Tax law makes it easier to put capital in than to retrieve it, like putting data into a bad filing system.

The Canadian tax treatment embodies the idea that the rents of Mine A are not really income at all if only they are sunk into Mine B. This entails two dangerous fallacies. First, it invites consolidation of accounts, loss of identities, and cross-subsidization, all of which are maleficent devices for having economic winners carry subeconomic losers, subsidizing the investment of scarce capital into ventures each of which, individually considered, would not pay. In the case of exploration the abuse is redoubled by the pooling of risk, which lends itself to the same fallacy, simply by pooling good and bad risks together on the ground that they are all "risks." Second, capital outlays are treated as current expenses.

Some mineral spokesmen allege they require the rent from low-cost units (often collected in the form of risk premia included in "required" rates of return) to cover losses elsewhere and exploration costs. If these rents were indeed all so expended it would dissipate the entire rent, meaning all labour and capital might as well be applied to other resources. There is not much point in that, especially for the Crown, which is left holding an empty sack. Truly, each individual expenditure should justify itself and stand on its own feet with regard to its probability of success. Any regression from this principle invites cross-subsidy and dissipation. Pooling has its uses, but when it degenerates into waste, it becomes an abuse.

After one finds a deposit or a field, and before one produces it, it normally appreciates. Accrual of value is not taxable before realization—a provision that in effect lowers the effective tax rate on all assets that appreciate over some years before sale. It is the deferral of tax that lowers the effective rate, regardless of other provisions. But in addition there is a depletion allowance so that part of the cash flow of producing deposits is untaxed. This allowance is limited, since 6 May 1974, to 25 per cent of production income, a less generous allowance than that in the United States, but still one not limited to costs. In Canada bonus bids and other land leasehold acquisition costs are deductible—in the States they are not, if depletion is chosen. To deduct the costs of acquiring and building up an asset, and then besides to deduct 25 per cent of its income, entails substantial double counting.

Property Tax Policy

Capital sunk into exploration on crown lands creates values that are largely free of property taxes. The same capital invested in city buildings would have to earn interest plus property taxes each year to earn its keep. But dry holes are exempt, as are geological maps and files of costly secret information. Possessory interests and occupancy interests are virtually exempt, as are access roads and rights of way. Marginal feeder pipelines are undertaxed, because assessments in British Columbia are based on load factors, which get lower toward the ends of any system. Most of the capital and land value used in and resulting from exploration and development on crown land simply is not subject to property taxes.

Minerals on private land are usually underassessed for property tax, but minerals on public land are not assessed at all. As crown lands are generally more remote, this is a decentralizing force. The more remote the lands are from population, the less likelihood there is of their being subject to future school taxes.

It has been alleged that a property tax on mineral reserves (such as Saskatchewan now levies on potash, Alaska on oil, British Columbia on sand and gravel, and most American states weakly on all minerals) will overaccelerate recovery. But the allegation is based on a partial analysis which ignores the fact that a property tax falls on competing investments. The equimarginal arguments of general equilibrium analysis require that any investment in deferral of extraction pay the same as other investments. Failure to tax mineral reserves results in over*de*celerated recovery from rentable deposits; failure to tax capital in marginal mines results in overdecentralized development.

Claimstaking

The rule of capture or "finders keepers" that pervades mining psychology and tradition and institutions contains a strong decentralist bias. It is a landgrabbing business not far beneath the patina of legal procedure, and the legal rules reflect this underlying spirit. Resources firmly under control may be held in reserve. The preternaturally frantic urgency is to stake out new treasures before those greedy other fellows. The more precarious tenures get priority.

Thus the United States federal government has held much of the Gulf of Mexico back from leasing, while its foreign tax credit and military-C.I.A. support boosted exploration in other nations. For a long time, too, the United States held Canada in reserve, abetted by Alberta's restrictive prorate policy. Canada, in turn, has heavily subsidized subeconomic scouting in the high Arctic, where the nuclear submarine wolfpacks rove— the national dominion is precarious and urgently requires confirmation.

Observe oil drilling in the North Sea. The big strikes are almost on the boundaries between the preserves claimed by the neighbouring nations.[7] Drilling occurs at the boundaries first; territoriality is the name of the game. Boundaries are always precarious, and unratified boundaries doubly so. They attract explorers like flies around a pot of honey.

Sliding Scale Royalties

Sometimes we find higher royalties applied to bigger wells or mines. British Columbia and Alberta both apply this rule to oil royalties. The evident presumption is that large scale development is identical with low unit cost. That is a half-truth and leads to capricious or random results. It is another form of tin can surgery.

To the extent that the preceding presumption *is* true, its rationale is faulty. Low-cost units have high-cost fringes, as noted earlier, and high royalties cause high grading. High grading and slow grading of rentable units coupled with low grading and early use of subeconomic units means overdecentralization.

Internalizing Earnings

Suppose the Crown avoids enough pitfalls eventually to collect some rent. What happens next? There is another snare before the public can benefit, and it relates to the agency in charge. Many officials in charge of any resource that yields revenues develop appetites to internalize earnings and build the industry, the professional fraternity, and the power base. They develop in-group, in-house ideologies (complete with industrial dependencies and constituencies) that tell them it is immoral for any dollar earned in mining (or forestry, gas, transportation, or power, among others) to be "diverted" to other uses. This is familiar bureaucratic behaviour. The antidiversion provisions require that the captive capital be reinvested in the agency empire, and that means overdecentralization again.

The British Columbia Petroleum Corporation is an example. To get the premium price for "old gas," producers must now reinvest the revenue in exploration and development. Saskatchewan law and federal law have the same effect via tax policy.

Cheap Money

When governments decide to foster something they usually subsidize the capital input primarily—they find ways to apply cheap money rather than cheap labour. We have seen how tax policy acts to force cheap money into exploration and early development of new mines; we have seen how empire building officials seek to internalize profits and supply their agencies with cheap captive capital. In addition, it is common for governments to lease valuable lands on easy terms to large firms in order to improve the firms' credit ratings—to help them raise money at lowest rates. This is a perennial plea of lessees seeking easy terms, and governments often go along with it.

Cheap money in resource economics is another decentralizing bias. The effect of cheap money is to attract and hold capital in mining. That means earlier inception of new projects, longer planned programmes of development and schedules of extraction, and slower liquidation of established deposits. The effect is exactly the same as exempting mineral deposits and capital from property taxes. In short, the availability of cheap money makes the industry carry too much fat and spread itself out like a person who overeats and underexerts.

The use of cheap money in mining has philosophical support from many persons who wish to expiate their guilt toward future generations. The idea is that cheap money means less discounting of the future and slower extraction. This ignores many points, but here we need only note that it is self-defeating. It is true that cheap money in mining is going to slow the use of on-stream deposits but it speeds the development of new ones. Overall it must increase total current output because it draws more capital into mining. Each mine may produce more slowly, but new capital must be attracted to marginal opportunities until the advantage of cheap money is offset by lower prices and higher costs.

Valorization

The mineral industries are marked by cartel schemes. These usually involve the active cooperation of governments with private conspiracies in restraint of trade—save that when government lays on its hands, "conspiracy" becomes "orderly marketing." Cartel restraint takes the form of partly shutting in and slowing down the superior, low-cost mines.

Success in valorization means the seller nets some monopoly rent on top of other rent. What then will the seller do with the flow of cash? Many things, to be sure, are done, but one frequent answer is to use it to underwrite marginal exploration ahead of its time. This squirrels away capital for the future with minimal immediate impact on price. But over time the new deposits get quotas, too, resulting in overdecentralization.

All the foregoing factors add up to a powerful bias to overdecentralize mineral extraction, and thus are to be eschewed and/or compensated for in administering crown lands for the general good.

OVERDELEGATION

The Crown should avoid counterproductive meddling on the one hand, but it should also avoid the other extreme of delegating too much authority to one giant lessee, one chosen instrument—like the old East India Company or Hudson's Bay Company or the Canadian Pacific Railway—with the thought that it can "internalize externalities" and so achieve economies of scale in land development. Overdelegation is dereliction. We, the government, *are* the giant firm to whom this has been turned over. It is our job to create the institutions and infrastructure within which atomistic private enterprise can flourish; it is not our job to let some privileged private parties create institutions and infrastructure. They would internalize us.

The only way to find rent is to have competition for land in a free market. One giant lessee, even if efficient, will not generate this information. There must be an active market for crown lands. The Crown will be guided towards formulating the right policies simply by performing those functions prerequisite to having a viable market for its lands. To perform this function adequately is challenge enough for any administrator. It will occupy his time quite fully and very usefully.

Initially the Crown should avoid delegating the timing of issuing permits and leases. There is an optimal time for issue, from the Crown's viewpoint —the time when the present value of anticipated revenues stops growing faster than money in the bank. Canadian federal policies have let industry carry the initiative and tie up most of the North West Territories and offshore acreage far too soon. If the Crown were simply to employ people competent to decide on the timing of initial bargaining with permittees and lessees, the Crown would go far toward preparing for needed later functions.

The Crown should not let lessees dictate the size of units put up for bid. The more respectable, gentlemanly, cultivated, and articulate majors will naturally speak knowingly and persuasively of economies of scale, financial responsibility, pooling of risks, marketing contacts, lines of credit, and so on. That is their interest. The Crown should protect ours and, like a real estate subdivider, carve up the land to maximize the value. If the Crown should err, let it be on the small side, to foster large numbers of smaller competitive firms.

OVERALLOWANCE FOR RISK

Overallowance for alleged risk is an easy way to pad capital costs; it sounds so innocent and reasonable and legal. But it converts our rent into their income.

Are risk premia required in mining? Alfred Marshall wrote, "If an occupation offers a few extremely high prizes, its attractiveness is increased out of all proportion to their aggregate value." Lotteries, sweepstakes, casinos, and numbers games yield great returns almost wherever permitted to operate. Investors buy growth and glamour stocks and other speculations at very high P/E ratios. While some people prefer the steady yield, clearly there is a substantial fringe of others with a taste for risk and the associated *macho*. Some choose crime; some gambling; some growth stocks; and some prospecting. Some allege that more money goes into mining than ever comes out; others merely say the overall rate of return has been low, historically. Neither view jibes with the claim that capital requires high risk premia to enter mining.

Sometimes "risk premium" is a clumsy way of saying investors must make enough from the winners to return something on the capital in the losers, as well. That is fair enough but is quite different from and much more modest than claiming a risk premium on the total investment in losers plus winners. If the pleader wished to communicate clearly he could distinguish between a "reserve for dry holes" and a true risk premium. The former clearly has a rationale; the latter has not, while Las Vegas thrives. The "dry-hole reserve" is not income, it must go to cover losses. The true risk premium is income, and it may be removed and consumed before there is any return of principal invested. It covers no losses at all; it merely rewards investors for playing in a game where there are losses that must be covered —some other way.

One often hears that investors "require" or "target" a rate of return at some high level like 20 per cent, or they will not play. At the same time one hears that historical returns have been very low on total outlays (just as they are in gambling casinos). To reconcile these conflicting claims we have to conclude that the "required" rate of return is really some vague expression of what one hopes may happen if all goes well, before he will put it to the touch and take the chance of losing something. "We require 20 per cent" gives a spurious air of precision and brainwork to accounting in a very chancy business where actual returns are lower than 20 per cent.

As for the risk of losing your ante, the risk premium itself raises that risk and necessitates itself, a very circular kind of reasoning. If a man says "I never got back the money I sunk in that hole (I just earned 20 per cent on it for 25 years)," he is greatly exaggerating his woes. Earning 20 per cent for 25 years without recovering principal is the same as earning 19.78 per cent with complete recovery of principal—yes, really, it can be checked out in a set of interest tables or on an HP-80. All one need do is increase that risk premium the tiniest amount to be able to say he lost his stake—thus justifying the risk premium in a circle of reasoning he can extend as far as he dares.

It is a good talking point because most people are unaware of the quantitative relationships of interest rates and amortization. Politicians and publicists learn to throw around terms like "profits" and "risk" and "payback" as though they knew what they were talking about, but only the odd moneylender stops to define terms or look at the numbers. So the artful mining lobbyist presents his case to the best advantage. If he said, "We only make 19.78 per cent on Sourdough Mines when our target rate is 20 per cent," he would weep alone as his listeners mocked and jeered. So instead he says "We aren't getting any payback of our investment," which sounds much more drastic, and he may win some relief.

A reserve for dry holes has more rationale, as noted earlier. This gets into the pooling of risks, which is valid up to a point—the point where we pool good and bad risks. Then we are engaged in cross-subsidy, and it is time to ask "Who needs dry holes?" They are a means, not an end. Each exploration needs to justify itself in terms of its own individual probability of success. Each success needs to yield enough to cover *some* failures. The question is "how many failures?" The risk premium approach opens the door to raids on crown rents far in excess of legitimate reserves for dry holes, and it converts a reserve for losses, which is in no way income, into investor income to take home and spend for fun.

OVERADMISSION OF PROSPECTORS

It has become a cliché of modern resource economics that open access to common grounds permits overuse. New entrants crowd onto the open range so long as there is any return above cost—rent, that is—until all rent is dissipated. This doctrine has its classical expression in Arthur Young's "The magic of property turns sand into gold"[8] and its contemporary revival in Garrett Hardin's *The Tragedy of the Commons*.[9]

This principle, which is fairly obviously applicable respecting fisheries, grassland, beaches, and streets, is equally applicable, but less obviously so, with respect to prospecting. Under British Columbia's Free Mining laws there are free-ranging permits to explore. The "free miner" may enter onto all lands of the Crown and upon any other lands where the mineral rights are reserved to the Crown to stake claims and produce. In the United States the

Mineral Leasing Act of 1920 permits anyone to prospect on federal lands.
Canadian federal oil and gas policy allows nonexclusive exploratory
licences.[10] Exploration cost is small relative to other mining costs, but it
comes earlier, and the unrecovered capital claims payment of interest. To
earn 10 per cent it must double every seven years. So if exploration
antecedes development by forty-two years, one dollar spent on exploration
costs as much as $64 of investment in development. Thus premature
investment is a form of overinvestment. It piles up a huge burden of
unrecovered capital which later income must repay.

The explorer, however, will try to stake a claim just as soon as he thinks the
claim is worth as much as his outlay. He is the fisherman and the claim is
the fish. The claim has a present value long before the optimal time to begin
producing. The value is evanescent because someone else may get there
first. So as soon as the present value equals probable finding costs, out go
prospectors to stake claims. Compound interest on premature outlays then
may well eat up all the rent.

There is a time when mineral land should be swarming with and
supporting many people—the time when it is ripe for use. But there is an
earlier time when land is best held in reserve. To grubstake premature
prospecting is to overapply expensive capital to land and let financing
consume the potential rent.

UNDERPRICING

A common way of sharing mineral rents widely is to underprice the
product. The United States Federal Power Commission has done this for
years with natural gas. In Canada, an exporting nation, no one wants to
share the rent abroad, but sellers of energy are forced to share with domestic
consumers. This results in a two-price system. The nation has one such
system for oil, and British Columbia has another one for gas. Domestics buy
cheap, and foreigners dear.

There are several faults in the system. One is resource waste by
consumers. To forego a world oil price of $10.50 a barrel by selling oil
domestically at $6.50 a barrel is just as wasteful as buying at $10.50 to sell at
$6.50.

And who is this worthy domestic consumer we subsidize? Energy is pretty
clearly a superior good. It is the wealthy who own big homes to heat and
have big cars to fill. The firms using most energy are those which have gone
furthest substituting machinery and other capital for labour; farmers with
the most land to till; and mineral refiners whose payrolls are a small share
of the value added. Burning energy creates noise and other pollution. Using
our not-so-progressive tax system to subsidize wealthy people who waste

energy and pollute the homes of people who can't afford to buy the scarce unpolluted land is as regressive as it is wasteful.

An even more common kind of double pricing is the rebate granted to domestic processors. British Columbia's Mineral Royalties Act (Bill 31, 1974) contains an abatement of royalty for concentrates shipped to domestic processing plants. This is a way of supplying feedstock to local firms below the world price. The idea is to share rent with labour by creating jobs.

To create jobs, however, we must lower the cost of hiring labour. Cheap feedstock encourages the substitution of feedstock for labour, a substitution for which there is considerable scope. Rather than a rebate on mineral tax, what is needed is a rebate on wage taxes.

While underpricing fosters wasteful consumption, it simultaneously cuts down production. The British Columbia Petroleum Corporation made considerable profit the first year it assumed a monopoly of marketing gas for export. Its notion is to collect rent by buying cheap in the field and selling dear at the border. The trouble is, of course, that a uniform low field price fails to distinguish low-cost from high-cost producers—it treats them all like low-cost producers and suppresses half the supply.

So in its second year the Corporation was forced to raise field prices to raise supply. Logically this would soon lead to a nonprofit marketing agency setting a market-clearing price. That is, it leads to the other extreme, the one the industry constantly promotes, of treating all like high-cost producers. The government cannot let that happen, however, and one hopes that the government will soon look for better ways to collect rent.

The lesson to learn is that rent arises because the earth is not uniform. Low-cost gas can sell for as much as high-cost gas, so low-cost gas yields rent. The only way to collect rent is at the source, by identifying the low-cost deposits and charging higher rents for them.

The kind of thinking that obstructs clear understanding of that simple, basic, and essential point is exemplified in the practice of equating "rent" and "windfall." Windfall results from unexpected luck like the rise of demand for our gas exports. It adds to rent, but it is not the whole of rent. Rent exists because land is finite, and some land is better than other land. People pay rent for better land because they can produce at less cost per unit of output. Such reasoning is simple and obvious, yet many people overlook it and do great damage as a result.

Two British Columbia mining companies have reported a net income equal to half their sales.[11] That is one indication of a low-cost, rent yielding mineral deposit. Many others report little or no net income at all. It is of considerable interest in a society threatened by overconcentration of wealth and economic power that the rent collecting firms average out much larger than the marginal ones. Devices like general underpricing fail to distinguish the fat from the lean and fail to collect rent. At the same time they miss a splendid opportunity to strike at the core of concentration of wealth and market power.

CONFUSION OF RENT AND PROFIT

Some analysts refer interchangeably to the "profit" and the "rent" of mines. The terms do not mean the same thing, however, and this careless usage can cause confusion and error.

The "profit" from a property is the sum of the true rent plus the returns imputed to the capital used to develop it. In common accounting terms, profit equals cash flow less only depreciation.

To find the rent we must deduct, in addition to depreciation, the income of capital. This is the interest on unrecovered capital. Depreciation and income of capital together are called "capital recovery". Capital recovery is the figure shown in Table 1, column 2. It is the percentage a debtor would pay annually to retire a debt while also paying interest on the unretired balance. In the present analysis (as in all general economic theory), the term "interest" refers to the return to capital, whether or not there is explicit contractual interest paid on a debt.

Compactly:

$$\text{Cash Flow} = [\text{Gross Receipts}] - [\text{Current and Ancillary Expenses}] \quad (1)$$
$$\text{Profit} = [\text{Cash Flow}] - [\text{Depreciation}] \quad (2)$$
$$\text{Rent} = [\text{Cash Flow}] - [\text{Capital Recovery}] \quad (3)$$
$$\text{Capital Recovery} = [\text{Depreciation}] + [\text{Interest}] \quad (4)$$
$$\text{Rent} = [\text{Profit}] - [\text{Interest}] \quad (5)$$

In a marginal mine, cash flow equals capital recovery, profit equals interest, and so rent equals zero. However, the profit income going to capital may still be large. We have already stressed in Table 1 and elsewhere that capital recovery is many times depreciation whenever capital recovery is slow. In mining, capital recovery is often slow so the excess of profit income over rent may be substantial.

The statesman who attempts to socialize all the profit of mine property would drive all the capital out of mining and be left collecting only so much rent as labour could generate absent capital. Now it is true we have made mining overly capital-intensive as already noted, and there is room for some countermeasures. But to drive all capital out of mining is clearly far beyond the proper goal.

The result of trying to socialize the profit of mine property therefore would be the realization that it cannot be done. The government would then retreat to a lower rate, collecting less than all the rent from low-cost mines and aborting much productive investment. To avoid this doubly unhappy outcome, we need to distinguish rent sharply from profit or income, as shown.

Another important difference between rent and profit corresponds to the legal difference between a claim *in rem* (against the thing) and *in personam* (against the person). Rent is a claim *in rem* against the mineral deposit, regardless of who the lessee may be and regardless of his circumstances.

Profit, on the other hand, imputes to persons (including corporations). If they are in debt, profit is a personal concept, net of contractual interest paid to lenders and may be as low as zero, indeed lower. They may consolidate accounts and allocate large overhead charges to any given mine. They may even invest in exploration elsewhere and claim it as an expense against a local mine. They may allocate cash flow to reserve accounts and disclaim it as income. They may transfer profit elsewhere by buying too high and selling too cheap. In short, they may use the whole bag of tricks which multinationals use to outdo natives.

The advantage to the Crown of tapping rent rather than profit is evident. The Crown need only consider the circumstances of the mine alone, and one mine at a time. Losses on other mines and businesses are not our concern, nor is the overhead of the New York or Toronto headquarters. [12]

There is another legal aspect of rent which is of immense importance to the province. The British North America Act clearly sets aside the income of lands as a provincial preserve. That means rent not commingled with interest or wages which Ottawa may tax. It is by failing to define and insist on the distinction between rent and other income that the province has played into Ottawa's hands and made provincial revenues vulnerable to raiding.

The armour of the western provinces (where crown ownership of minerals is most significant) is section 125 of the B.N.A. Act: "No Lands or Property belonging to Canada or any Province shall be liable to taxation." The section read apart from its context might be interpreted very broadly, but has not been. In *Attorney-General of British Columbia* v. *Attorney-General of Canada*, [13] section 125 did not override the right of Ottawa to tax liquor imports of a provincial marketing agency, for example. The rather tangled line of judicial reasoning is expressed by Professor Gerard La Forest, in part, as follows: "section 125 does not provide a general immunity from taxation to the Dominion and provinces. Only their 'lands' and 'property' are exempted. . . . section 125 was intended merely to prevent the use of the taxing powers by either the federal or provincial governments in a manner that might impair the control of the other over its property; the taxes imposed here [on liquor in B.C.] were justifiable under the trade and commerce clause which. . . were not affected by section 125." Section 125 deals with "not taxation in general, but the liability of property to taxation." [14]

Finance Minister Turner made clear his position that provincial royalties had taken on the character of "disguised taxes." The provinces held that they were "reservations," an incident of property, and not taxes. Yet it is hard to maintain that a general royalty is an incident of property when it is collected at the same rate from production on valueless as on rich rentable property. It is not related to land rent or value so closely as to labour and capital inputs. That the federal view prevailed reveals the weakness of commingling rent with wages in the royalty base. Royalties are not based on land value or land income but on gross activity, an even broader base than property income. They proved easily raidable. Marketing agency profits are just like royalties, and they proved raidable. But property taxes and stumpage, which get closer to being direct taxes limited to land income, have not been raided. The province must disaggregate and impute rent to save its own revenues. Economics is the art that shows how.

In the unlikely event that section 125 should be cast aside completely and the issue turned over entirely to current electoral results, there is another and nearly invulnerable defence line, one that is also good economics and good administration. This is to alienate crown lands, selling the fees for a good price to the highest bidders, but subjecting them to property taxes at a high rate. Property taxes are certainly deductible, by law and long custom. The government that disallowed their deductibility would have to answer to massive electoral powers in all ten provinces.

The province that regards alienation as taboo locks itself out of this option. That is a mistake. Taxation asserts the Crown equity in land just as surely as do rents and royalties, and there is plenty of history to show that private taxpaying holders of fee simple title can manage property as well as lessees, occupiers, licensees and tenants of the Crown.

OVERLOOKING THE TAXATION OF NONMINING ACTIVITY

A good deal of industry pleading and some scholarly analysis proceed on the implicit presumption that other uses of labour, land, and capital are not taxed. If one overlooks that fact, it is easy to conclude that taxes, royalties, or other charges drive resources out of mining or distort decisions. But in fact there is only an antimining bias when mining is taxed more heavily than alternative resource uses.

Indeed, with that in mind, it might almost seem that a perfect system of collecting mining rents, if we could create one, would be in its very perfection imperfect, because it would draw too much labour and capital into mining. The taxes on other uses of labour and capital are clumsy and onerous, and if we should collect rent without the excess burden of aborting marginal increments of labour and capital, we would destroy the "second-best" balance and overintensify mining land.

It is a dreary doctrine which bids us endeavour for mediocrity and scorn perfection. Fortunately it is false. The reason is that one province is a small part of the world. It is true that a perfect system of collecting mining rents would attract extra[15] labour and capital into mining, but the net extra labour and capital would come from outside the province. Each immigrant would improve his own lot and also generate some rents for the Crown. Even though we ended with mining overdeveloped relative to other industries in the province, we would all be better off.

Other provinces may lose tax or other revenue from the marginal people lured our way. However, we have no control over other provincial policies; we can only presume they are doing what they think best for them. They always have the option of emulating our policies if they think them in their interest.

There may be proper applications of "second-best" reasoning, however. One concerns allocating land among competing uses. A province can draw on the world capital pool and national labour pool, but its pool of land is fixed, so the foregoing lines of argument do not apply to land. If we achieve a perfect tax or rent collecting policy in mining, but not in forestry, we will lure land from forestry into mining. There is only limited competition for land between the two, but there is greater competition between mining and recreation and agriculture. As we aspire to improve mining policies, we therefore need look to the margins of competition for land and improve policies in other industries. Alternatively we might strive for uniform suppressiveness in all industries and drive people and capital out of the province. Antigrowth forces may prefer this course, but one may doubt that the people who drive are morally superior to those who are driven. In any event, to achieve uniform suppressiveness is not practical.

A kind of partial equilibrium argument commonly made against royalties is that they cause miners to high-grade deposits. There is much truth in that, but it is much too partial. There are also many subsidies that encourage low grading. One is the depletion allowance in both federal and provincial income taxation, which is a negative royalty. The rate is high compared to most positive royalty rates. To make a big issue of royalty induced high grading while overlooking the larger and opposite effects of the depletion allowance would be misleading. The greater problem with royalties is that they substitue for cash delay rentals and bonus bids, and cause lessees to hold too much good land too cheaply without pressure to use it.

Another subsidy is in transportation. Freight rate schedules nearly everywhere follow a "value-of-service" rather than "cost-of-service" principle, resulting in a large cross-subsidy favouring primary producers. Secondary and tertiary producers whose goods have a high labour content carry the cost for primary goods of high land content. Like the depletion allowance the subsidy varies as a function of output.[16]

Another encouragement to low grading is the complex of tax subsidies and other provisions treated herein which pump cheap capital into mining. Some inputs complementary to capital (energy, for example) are subsidized too.

What is not subsidized in mining is the labour input. The problem is not a lack of encouragement to mining, but unbalanced encouragement. The net result of the various forces is not a bias against mining, but a bias against labour. Mining uses too much capital, too much land, and is undermanned.

The bias against labour is not peculiar to mining; it continues down to the consumer. Payroll taxes pervade all of economic life ubiquitously and are collected in advance with an iron and unforgiving hand.

One cannot study our tax laws without concluding that Parliament regards giving and accepting employment to be a social evil, like dealing in liquor or tobacco, to be discouraged by heavy discriminatory taxes coupled with rewards for abstinence. Payroll taxes on miners do not therefore drive labour out of mining peculiarly. Rather, payroll taxes everywhere cause employers to prefer capital, and potential employees to prefer welfare.

Coupled with minimum wage laws and union wage scales, payroll taxes cause employers to high-grade the labour force, leaving marginal workers on the bench, on unemployment insurance, in early retirement, in extended adolescence, in police blotters, on frequent strikes, in doll's houses, or other forms of involuntary idleness. This is the kind of high grading that the total tax system induces. This is a true social problem and a giant one, created by inept policy. The other kind of high grading is something of a figment by comparison.

In a world of heavy payroll taxes, what is the effect of net subsidies to mining? Downstream of mining there is great scope to substitute feedstock and energy for labour—to produce goods of higher resource content and less labour content. That is what the world has been doing; using lots of resources and high grading its labour. Net subsidies to mining helped to cause that. They make raw material feedstocks too cheap relative to labour.

Surveying the total system, royalties do less overall damage than a partial analysis would suggest. They still rank low among policy choices, but analyzing the total system forces us to put a higher priority on abating the system's overall bias towards substitution of primary feedstock for labour.

OVERCONSOLIDATION OF ACCOUNTS

We have mentioned this subject before in connection with underpricing. The strong use the weak as front men—or front widows, often enough. Holders of low-cost deposits pass off high-cost marginal deposits as typical, so all income of property may be regarded as functional and necessary. The way to beat this game is to disaggregate, to analyze each deposit separately.

A firm or public agency with many deposits, some rentable and some poor, consolidates its books so the rents are soaked up by the costs of the poor. This is especially worthwhile for the firms if the "poor" are merely premature, and capital outlays may be passed off as current expenses. Consolidation of accounts lets the rich look poor to avoid taxes and lets the unfit look viable to justify their continued existence.

Again the answer is to disaggregate. Large firms will attempt to justify their keeping high rents on the grounds of losses taken elsewhere, as though losing money were a social service. The objective of a government leasing lands should be to collect all the rent from rich lands and let the poor ones go. There is no call for any cross-subsidizing; and if there were, there is no private corporation with the moral authority to do it for us.

CONCLUSION

There are, then, eight common errors to avoid in the effort to maximize and collect rents from crown lands: overdecentralization, overdelegation, overallowance for risk, overadmission of prospectors, underpricing of primary products, confusion of rent and profit, overlooking of nonmining taxes, and overconsolidation of accounts. In reviewing them one appreciates more fully the proverb "It is easier to face a common enemy than to share a surplus." There are so many inane ideas about how to share a surplus that it is no wonder many polities cannot stand prosperity. Toynbee's *A Study of History* suggests it has not been the common enemies but the ineptly shared surpluses that destroyed great civilizations.

This article has attempted to clear ground, to dispose of fallacies and diversions so that we might focus on the substantial real problem of how to collect rent from crown lands. While the writer chafes to address that topic in a positive way, limits of space preclude it here. Prudence, too, might suggest that he defer to his coauthors who are so well equipped to do the job.

Notes

1. This formula is used to calculate *allowable cut* from public forests in Canada and the United States. It says that the annual cut should be no greater than the mature stock divided by the putative future rotation age (usually overstated) plus current annual increment.

2. James R. Nelson, "Energy Industry Alternatives for the Future," in John W. Wilson, ed., *The Energy Industry* (forthcoming).

3. D. Gale Johnson, "Resource Allocation under Share Contract," *Journal of Political Economy* 58 (April 1950):111-23.

4. Harvey Averch and Leland Johnson, "Behavior of the Firm Under Regulatory Constraint," *American Economic Review* 52 (December 1962): 1052-69.

5. MER stands for "Maximum Efficient Rate," but there is nothing efficient about it. It is a transparent rationalization for slow production by ignoring carrying costs and has no respectable support among economists. It is a remarkable example of how insider technicians, by calling something "efficient," can intimidate the public into accepting it.

6. For a detailed history of "market demand" prorationing in Alberta and of the Alberta system generally, see G.C. Watkins, "Proration and the Economics of Oil Reservoir Development" (Ph.D. diss., University of Leeds, 1971) and "Regulation and Economic Efficiency," Discussion Papers Series no. 32(Calgary: Department of Economics, University of Calgary, 1975), pp. 4-9.

7. See *North Sea Oil and Gas,* Study sponsored by the Council on Environmental Quality (Norman: University of Oklahoma Press, 1974), pp. 11-28.

8. Arthur Young, *Travels in France,* Vol. 1 (1790), p. 88.

9. Garrett Hardin, "The Tragedy of the Commons," *Science* 162 (December 1968): 1243-48

10. Michael Crommelin, "Offshore Oil and Gas Rights," *The Natural Resources Journal* 14 (October 1974): 463-72. The American "permit" and the Canadian "exploratory license" are nonexclusive as to location.

11. The firms are Lornex, a Noranda subsidiary, and Placer Development, a Rio Tinto Zinc(RTZ) subsidiary. RTZ also owns Bougainville Copper, Ltd., which, in 1974, made $236 million on $372 million in sales, or 63 per cent. It has been claimed that Rainbow oil from Alberta "costs less than 10¢ a barrel to find and produce": see *Business Week* (1 July 1967), p. 71. We believe 10c is below the correct figure, but not by much.

12. If in the process we create a small bias in favour of resident owners, there are plenty of biases against them that want offsetting. Fostering local ownership is consistent with public policy and in the interests of the province. Indeed it is undoubtedly in the interest of the whole world to strengthen small, decentralized, atomistic enterprise against the octopi of high finance and absentee control.

13. [1924] A.C. 222

14. G.V. La Forest, *The Allocation of Taxing Power under the Canadian Constitution* (Toronto: Canadian Tax Foundation, 1967), pp. 150-53.

15. Not extra relative to now, but relative to a hypothetical system of uniformly burdensome taxes on all industry.

16. Capacity-to-serve is a fixed subsidy, as well. The annual carrying cost of track serving intermittent miners is carried by the steadier parts of the system.

An Economic Theory of Mineral Leasing

FREDERICK M. PETERSON

In this article an economic theory of mineral leasing is developed and used to formulate and evaluate leasing policies. Minerals are understood to include oil, gas, and coal, as well as hardrock minerals. I consider optimal exploitation and leasing under ideal conditions, then relax assumptions to discuss the uncertainty, externalities, and market imperfections relevant to British Columbia. In the ideal model, the government can lease, rent, or sell a piece of mineral property in a number of ways, some of which have not been discussed elsewhere. The market works efficiently; so the government can use any policy or combination of policies necessary and recover the full present value of the property, providing the time path of mineral exploitation is not disturbed. Any policies that encourage or discourage exploitation or disrupt the pattern of activities are bad. These include depletion allowances, chequerboard leasing, and exploratory work commitments, since they encourage exploitation or change the pattern of exploration and development. When more realistic assumptions are considered, the market solution loses some of its appeal, the method and pattern of mineral leasing ceases to be arbitrary, and some distortionary policies acquire the attribute of offsetting other distortions in a second-best world.

AN IDEAL MODEL

Mineral properties controlled by the Crown are partitioned into n tracts, each with its own characteristics. At this point, the tracts have no use other than mineral extraction, and they contain only a single mineral. Such characteristics of the tracts as location, size, and quality of mineral deposits are known, so the deposits need only be developed and extracted—exploration is not necessary. Each tract has a cost function, $C_i(\dot{X}_i, X_i, t)$, giving the dollar per year rate of capital and operating expenditure on mineral exploitation as a function of the extraction rate in tons per year, the cumulative amount of extraction in tons, and time, respectively. Costs, which are known, reflect the tract's characteristics and the current state of technology. Costs on one tract are not affected by extraction on other tracts, so common pool problems are not present. It is assumed that marginal cost, $\dfrac{\partial C_i}{\partial \dot{X}_i}$, is positive and rising for high rates of output, to reflect diseconomies of crowding near the ore bodies. Costs are assumed to rise with cumulative output, reflecting the depletion of the better deposits, so $\frac{\partial C_i}{\partial X_i}$. is positive. At any point in time, the industry has a demand curve for the mineral which is represented by the function $P = (\dot{X}, t)$. $\dot{X}$ is the output of all tracts, both inside and outside the jurisdiction of the Crown. The Crown's holdings are assumed to be too small to affect the price of the mineral.

The Optimal Time Paths of Extraction

The Crown wants to set the rate of output for each tract, $\dot{X}_1$, $\dot{X}_2$, ... $\dot{X}_n$, so that the present value of each tract is maximized. As Hotelling and others have shown,[1] maximizing the present value of each tract is equivalent to maximizing the present value of consumers' plus producers' surplus for all tracts, so long as everyone is a price taker.

The present value at time zero of ith tract is given by the infinite integral

$$PV_i(0) = \int_0^\infty [P\dot{X}_i - C_i(\dot{X}_i, X_i, t)] \, e^{-rt} dt \,, \tag{1}$$

where r is the discount rate. The Crown maximizes the integral subject to the constraints that $\dot{X}_i$ be positive and that X_i not exceed the initial stock of minerals on the tract. Rather than find a single value of a variable to maximize a function, as in ordinary calculus, the Crown finds the whole time path of $\dot{X}_i$ to maximize a functional, a task accomplished with the Pontryagin Maximum Principle.[2] The Crown's solution of the maximization problem can be found in the Appendix. Only the results are presented in the main text.

In maximizing the present value rather than just current profits on a tract, the Crown takes into account the user cost, or the cost of depleting the resource,[3] as well as current extraction costs. The maximum present value of tract at time t, $PV_i^*(X_i(t), t)$, is a function of the amount of extraction that has occurred to date [see equation (15) in the Appendix], and marginal user cost, q_i, is defined as the amount by which this maximum present value falls in response to one unit of extraction $-\frac{\partial PV_i^*}{\partial X_i}$. For a maximum it is necessary that price minus marginal extraction cost minus marginal user cost equal zero,

$$P - \frac{\partial C_i}{\partial \dot{X}_i} - q_i = 0 \,. \tag{2}$$

In economic terms price equals marginal cost, where marginal cost includes both factor costs and use-related depreciation costs.

Each tract has an optimal time path of extraction that depends on its characteristics relative to other tracts and on market price. The best tracts are exploited first, and progressively inferior ones are exploited as the price of the mineral rises. A typical time path of extraction for a given tract is presented in Figure 1,

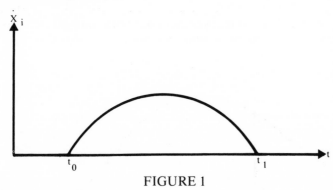

FIGURE 1

where t_0 and t_1 are the times when extraction begins and ends.[4] The time path for all the crown lands would be the summation of the individual time paths, as shown in Figure 2.

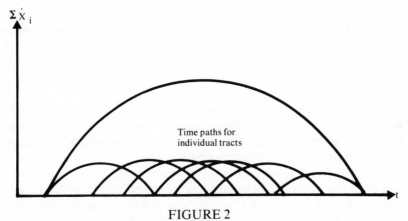

Time paths for individual tracts

FIGURE 2

The present value of a tract represents rents due to the site only. Any normal profits or rents due to mining expertise are included in costs and are therefore deducted. If $\dot{X}_i^*$ is the optimal time path of extraction, the rate of change of the present value of a tract is found by differentiating the present value integral with respect to t_s,

$$\frac{d}{dt} PV_i^* = - (P\dot{X}_i^* - C_i) + rPV_i^*, \qquad (3)$$

where C_i is a shorthand expression for the

cost function. When a tract is not being exploited and its revenues and costs are zero, its present value grows at the rate r like any other asset; the discount rate is the prevailing rate of return in the economy.[6] If its present value were growing any slower, the owner would sell or exploit it, while a faster rate of growth would attract investors and raise its current price until the growth rate fell to r. When a tract is being exploited, its present value levels off, falling to zero once everything has been removed. (It is assumed that the tracts have no other use.)

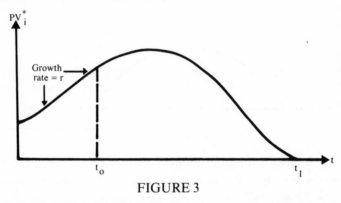

FIGURE 3

Note that the present value continues to rise for a while after extraction begins. This follows from the assumption that extraction starts up gradually, as it does in Figure 1, so that capital gains on the tract temporarily outpace its depletion.

Leasing of the Tracts

We have seen how the Crown could operate the tracts to maximize their present value. Rates of output would be set so that price equalled marginal extraction cost plus marginal user cost. Each tract would come into use and have an optimal time path of extraction according to its characteristics and market price. The Crown may want to recover the present value of the tracts while letting someone else perform the extraction. There are a number of options available, ranging from outright sale of the tracts to selling the mineral by the ton to a mining company. It is immaterial which course the Crown takes as long as the optimal time path of extraction is preserved. It is assumed that there are lots of prospective bidders for each tract, no matter what kind of arrangement is being made. If someone were to operate a tract owned by the Crown, he should have the incentive to follow the present value maximizing time path rather than some self-serving, short-run programme of his own. With one exception, each of the leasing arrangements incorporates this incentive.

a. Outright Sale. The Crown could sell the tracts to any number of buyers at the full present value. They would be satisfied because they could receive the prevailing rate of return before extraction began and the normal profits and rents afterwards. Since the tract has no other use, this means selling the mineral rights and amounts to selling perpetual mineral leases on a pure bonus basis. The characteristics and value of the tracts are known with certainty, so there would be no problem of risk or risk aversion. It would not matter whether the Crown sold a tract to an asset holder or a mining company, because a mining company would obtain the tract before commencing extraction. An asset holder would not delay extraction, because to do so would only decrease the present value of his tract.

b. Contract Mining. Getting the full present value on a tract is equivalent to getting the full profit flow, $P\dot{X}_i^* - C_i$, at all times—present value is just an integral of these profit flows discounted back to time zero. A mining firm could be contracted to extract the mineral according to a predetermined plan. The firm would turn over all revenues, PX_i, and be reimbursed for costs, C_i, which include the rents and normal profits due to mining activities. If not specifically instructed, the firm might change the extraction rate for its own purposes; consequently, the Crown would have to know the optimal extraction rate and enforce it on the firm. There is no other way to make the firm behave optimally without the built-in incentives found in the other leasing arrangements.

c. Profit Sharing. The Crown could go into partnership with a firm which would extract the minerals for a small percentage of the profits. The firm would behave optimally and maximize the present value of the tract because the rules for maximizing 1 per cent of the present value are the same as for maximizing 100 per cent.[7] In addition to the small percentage of the profits on the tract, the firm would also receive the rents and normal profits included in costs. On equity grounds, of course, one might be concerned whether 1 per cent or 100 per cent of profits went to corporations inside or outside the country. Profit sharing is used by several Middle East governments, which take up to 90 per cent of the profits on oil ventures.[8] Their calculated profits presumably include the rents and normal profits due to oil extraction firms; so 90 per cent of these Middle East profits could equal 100 per cent of rents due to government oil properties. In going into partnership, the Crown would have to know the business and monitor its partner, who might otherwise make unreasonable claims for costs or take other advantages.

d. Ground Rent. The Crown could capture the full profit stream and therefore the full present value of a tract by renting tracts for a two part charge. Rearranging equation (3), we see that the profits which flow from a tract can be divided into two parts, one a rental on the property, rPV_i^*, and the other a charge for depreciation, $-\dfrac{d}{dt} PV_i^*$,

$$P\dot{X}_i^* - C_i = rPV_i^* - \frac{d}{dt} PV_i^* \quad . \tag{4}$$

This is similar to the two part charge often paid when renting a car, where one part is a rental on the value of the car, and the other a mileage charge for wearing it out.

The current profits to the firm would be given by the equation

$$\Pi_i = P\dot{X}_i - C_i - rPV_i^* + \frac{d}{dt} PV_i^* \quad , \tag{5}$$

which would equal zero if the firm behaved optimally from the standpoint of the Crown. But could the firm do better by behaving in some other way? The answer is no. Since the maximum present value, PV_i^*, is a function of time, t, and the cumulative amount of extraction to date, X_i, its time rate of change can be found by total differentiation,

$$\frac{dPV_i^*}{dt} = \frac{\partial PV_i^*}{\partial X_i} \frac{dX_i}{dt} + \frac{\partial PV_i^*}{\partial t} = - q_i \dot{X}_i + \frac{\partial PV_i^*}{\partial t} \; ; \tag{6}$$

so by combining equations (5) and (6), the firm's current profits can be written as

$$\Pi_i = P\dot{X}_i - C_i - rPV_i^* - q_i \dot{X}_i + \frac{\partial PV_i^*}{\partial t} \quad . \tag{7}$$

To select the rate of output capable of maximizing its own profits, the firm would set the partial derivative with respect to X_i equal to zero,

$$\frac{\partial \Pi_i}{\partial \dot{X}_i} = P - \frac{\partial}{\partial \dot{X}_i} C_i - q_i = 0 \; , \tag{8}$$

and this is precisely the condition obeyed when maximizing the present value of a tract. The firm would take the long view because it had to pay for the depreciation of the resource. It would maximize the present value of the tract in the process of maximizing its own current profits. I am not aware of any rental arrangements like this being used for mineral resources. It would require the participants to know a lot about the resource and about future prices, so that the present value could be calculated on a daily basis.

e. The Severance Charge. The Crown could also require the lessee to pay by each ton extracted. If the cost per ton equaled the marginal user cost, q_i, the firm would behave optimally, but the Crown would have to charge an extra amount, M_i, in order to recover all the profits. The firm's current profits would be given by the expression

$$\Pi_i = P\dot{X}_i - C_i - q_i\dot{X}_i - M_i , \qquad (9)$$

which would be maximized by setting the partial derivative with respect to $\dot{X}_i$ equal to zero,

$$\frac{\partial\Pi_i}{\partial\dot{X}_i} = P - \frac{\partial}{\partial\dot{X}_i} C_i - q_i = 0 , \qquad (10)$$

just as a present value maximizer would do, since q_i equals marginal profits along the optimal path,

$$\Pi_i = P\dot{X}_i - C_i - \left(P - \frac{\partial}{\partial\dot{X}_i} C_i\right) \dot{X}_i - M_i , \qquad (11)$$

and we see that

$$M_i = \dot{X}_i \frac{\partial}{\partial\dot{X}_i} C_i - C_i \qquad (12)$$

in order for profits to vanish. When marginal costs equaled average costs, the extra charge would be zero. Generally M_i would be positive because marginal cost should exceed average cost when a stock resource is being exploited.[9]

The flow of payments, $M_i(t)$, could be paid at any time in a lump sum with the same present value. At time t, this sum can be expressed by the integral

$$B_i(t) = \int_t^\infty M_i(\tau)e^{-r(\tau-t)}d\tau , \qquad (13)$$

where τ is a dummy variable of integration. Paying a lump sum and a charge per ton resembles paying a bonus and a royalty, although it is doubtful that many royalties equal the marginal user cost or that many bonuses equal $B_i(t)$. From the Appendix, we see that q_i varies with time, in some cases growing and in others shrinking. It would be worth investigating the size and time path of q_i for actual mineral deposits. In all likelihood, q_i would not equal a fixed fraction times price as does the typical royalty per ton extracted. In the case of oil, one can argue that the royalty rate (and q_i) should drop to zero as exhaustion is approached to avoid the premature cutoff production.[10]

We see that any number of policies or combinations of policies can be used by the Crown to capture the present value of its mineral properties in this ideal model. Timing is not important. Tracts can be offered for lease or sale individually or all at once, and at any time, so long as resource extraction begins on time and the optimal time paths are not distorted. The market produces the present value maximizing time path of extraction for each tract. Any policy of the Crown that encourages or discourages exploitation or distorts this optimal behavior is bad. Policies under suspicion include leasing in chequerboard patterns, withholding regions from exploitation, setting minimum bids, offering exploitation incentives, requiring work commitments, or any other policies that affect the timing or pattern of exploration, development, or extraction of mineral resources. But we need to look at a more realistic model before condemning such policies.

MORE REALISTIC ASSUMPTIONS

Distortionary policies look better with relaxed assumptions about the availability of information, the number of bidders, the government's effect on mineral prices, the status of existing policies, and alternate uses of mineral properties. Under these new sets of assumptions, the market no longer assures the optimal timing of resource extraction or the maximization of the present value of government revenues, and the latter goal might no longer be optimal when political and environmental factors are considered. It is suggested that the government lease large tracts of land to correct an information spillover problem, and that the government become more involved in mineral exploitation to encourage the participation of smaller firms, to obtain a larger share of mineral rents, and to preserve environmental quality. A number of the arguments presented in this section draw upon my other article in this volume. "The Government Role in Mineral Exploration," which will be referred to by its short title, "Government Role."

The Availability of Information

The location and quality of mineral deposits is not known with certainty in British Columbia or any other place, nor can the future price of minerals or state of technology be forecast with any degree of accuracy. Instead of a world of perfect information, we have a world dominated by risk and uncertainty, and mineral exploitation is one of the most risky businesses, with costly exploratory programmes and low success ratios. When exploration is considered in the model, risk aversion becomes a problem, externalities appear, the number and size of firms change, and some leasing policies look better than others.

The ideal model could be modified by adding probabilities, dealing with expected values, and assuming risk neutrality. The conclusions about the efficiency of the market and the arbitrariness of leasing policies would not change,[11] but I think that risk neutrality is a bad assumption for the reasons discussed in "Government Role." Assuming that firms are risk averse to some degree and that exploratory ventures are risky, firms would be inhibited from undertaking some ventures that had favourable expected returns to the firm and the province. Small firms would be more inhibited than big ones because small firms have a higher probability of ruin. On lease sale bidding, firms would not bid the full expected value of a tract even if they knew that their expected value estimate was correct.[12] The government should alter its leasing policy in response to these problems. Outright sale of mineral rights and other lease transfer methods that involve big front end payments should be avoided. Profit sharing and royalty payments would be better because they shift some of the risk to the government and require firms to pay only if the ventures are successful. Also, the government would avoid the political embarrassment of having no share in extremely successful ventures. When left with too small a share of what turns out to be a very large pie, the government may be forced to reallocate the slices after the pie is being eaten, so to speak, and introduce political uncertainty—the worst kind of uncertainty for the firm. This may be the situation that British Columbia finds itself in today.

Large firms can diversify risks better than small firms, as Allais[13] and Slichter[14] have shown. If the probability of success is 10 per cent on a given venture, there is still a 35 per cent chance of total failure after ten ventures. Only a large firm would willingly face such odds, and the problem is exacerbated by the large amounts of money needed to find today's remote, hidden mineral deposits. As the easily accessible deposits are found by cheap surface geological techniques, large tracts have to be leased, sophisticated geophysical techniques used, and expensive operations conducted either offshore, in the Arctic, or at great depth.[15] As discussed in "Government Role," modern exploratory techniques have scale economies that may threaten the efficient atomistic competition of the ideal model, apart from the problem of risk aversion. Preston emphasized the importance of scale economies in nonferrous metal exploration,[16] and Shearer claimed to find a strong effect on the nationality and size of firms exploring for oil in Western Canada.[17]

If the provincial government wishes to favour Canadian firms, which are mostly small compared to the international giants, it should take the risk and scale economies problem into account. For instance, the government might perform some of the early exploration in virgin areas to reduce the risk for smaller firms. It was apparent on the North Slope of Alaska that the smaller firms joined in after the big risky wildcats had been drilled.[18]

Information spillovers also threaten the efficiency of the ideal model and

create biases in favour of large firms. Again as discussed in "Government
Role," the discovery of oil or copper on one piece of land provides
information about the geology of other pieces of land, probably miles away.
This is especially true for oil and gas as well as alluvial and stratabound
mineral deposits. The spillover effect spread over hundreds of square miles
on the North Slope of Alaska when the Atlantic Richfield Company made
its enormous discovery there, and lease prices skyrocketed. Even a dry hole
provides useful information about surrounding tracts. The prospectors and
wildcatters who provide this information often receive nothing for their
services. Surrounding leaseholders sometimes just sit back and wait for the
outcome of the exploratory efforts.

If the prospective mineral explorer cannot obtain a large lease block or
make agreements with the surrounding leaseholders, the project may be
delayed or cancelled even though its expected net benefits are positive. His
share of the benefits has to be great enough to cover his costs. All firms are
inhibited to some degree by the spillover problem, but the larger firms have
a better chance to internalize the spillover because they can afford extensive
leaseholdings. The government should offer large tracts for lease in virgin
areas and avoid claimstaking, chequerboard leasing, and lottery type
bidding, because such policies spread lease ownership widely and make it
difficult to assemble large blocks and arrange cooperative ventures. To
assist the smaller firms who cannot afford large blocks, the government
might, alternatively, act as a referee to force cooperation among small leaseholders or do the early exploration itself.

The Number of Bidders

In the ideal model, we assumed an infinite number of bidders so that any
number of tracts could be offered for lease at one time; but there is
considerable evidence that this situation does not actually obtain, including
what we have just learned about risk aversion, returns to scale, and
information spillovers. The larger the firms the smaller their numbers, given
the size of the mineral industry in British Columbia. Lack of bidders in
sufficient numbers could reduce provincial revenues on mineral leases and
could lead to various types of collusion.

There is a positive correlation between the number of bidders and the size
of bids on lease sales. Mead[19] found that oligopsony reduced government
revenue from the sale of Douglas fir timber, and Gaskins and Vann[20]
uncovered convincing evidence in bids for oil and gas leases on the Outer
Continental Shelf of the United States. In the latter study, the ratio of

winning bids to United States Geological Survey estimates of the value of the tracts rose with the number of bidders. On the March 1974 lease sale, the ratio rose from a curious 1.00 for one bidder up to 6.16 for eight or more bidders. The values do not necessarily imply collusion among the bidders,[21] but they do show a significant relationship between the number of bidders and the size revenues on lease sales. If the number of bidders can affect revenues by anywhere near the factor of six suggested by the Gaskins-Vann study, British Columbia should regulate the number of tracts and bidders involved in its lease sales very carefully.

There is probably a link between the number of tracts offered and the number of bidders per tract, because firms have budget constraints and capacity restrictions. They can only bid for and hold so many tracts at a time. They do not bid on twice as many just because that number are offered. This point is apparently understood by industry spokesmen, who worry about where companies are going to get the extra money to bid if the United States Department of the Interior drastically expands its offshore leasing program.[22] If the province increased the number of tracts, the bidders would spread themselves thinner and bid lower. The province could not offer all of its tracts at once without a loss of revenues, contrary to the suggestions of the ideal model.

Large firms and a correspondingly small number of bidders could reduce provincial revenues in another way. The firms could collude to decrease output and increase the world price of the mineral. They could purchase and hold mineral tracts in British Columbia to keep them out of production. World cartels are known to be important with many nonferrous metals. Decreased production would conserve resources, but provincial revenues would drop unless they were tied to cartel revenues. Consumers of the mineral certainly would suffer. Although bad or at least unnecessary in the ideal model, work commitments might be a good way to prevent a cartel leaving tracts idle and reducing government revenues.

Government Effect on Mineral Prices

In the ideal model, government holdings were too small to affect the price of the mineral, but the government may have monopoly power of its own. If the government could affect the price of the mineral, it could withhold tracts and reduce production so that its revenues were maximized. Producing company revenues might also be maximized. Although rationalized on conservation grounds, the prorationing programme of the Texas Railroad Commission effectively regulated prices and maximized state and company revenues. The United States government might control enough petroleum reserves to exercise some monopoly power; it was accused of increasing its oil revenues on the Outer Continental Shelf in the late 1960's by withholding tracts and setting high minimum bids. The government and the producers could both gain by such a strategy, depending on how the proceeds were divided, but the consumers would lose because of falling output and rising prices. When the consumers are foreigners whose goodwill is not sought, it might make sense to decrease output and maximize revenues, but one hesitates to recommend such a strategy when the consumers are constituents.

Existing Distortionary Policies

Distortionary policies were judged to be bad on the assumption that the world was perfect and no other distortions were built in; but many imperfections exist, and many distortionary laws have to be accepted as facts of life. It is possible that these distortions offset each other or that distortions can be cancelled by introducing other distortionary policies. Take the case of fixed percentage royalties, which discourage extraction and cause the premature shut down of producing mines and wells. Take also the case of gross depletion allowances, [23] which encourage extraction and cause the continued operation of mines and wells that should be shut down. Royalties used in British Columbia and elsewhere might be considered facts of life; so might gross depletion allowances, which are enjoyed by United States corporations operating throughout the world. Could the two distortionary policies cancel each other? As a matter of fact, a gross depletion allowance of the correct percentage exactly offsets the allocative effects of a royalty in the ideal model. When the corporate profits tax rate is 50 per cent, a depletion allowance offsets a royalty of the same percentage. [24] The present value maximizing time path of extraction is reestablished and mineral exploitation is efficient. Equity is another matter, which would have to be examined separately, but the point is that distortionary policies like percentage royalties and depletion allowances should not be rejected out of hand in a second-best world. They might be an improvement.

Alternate Uses of Mineral Properties

In the ideal model, a tract was useful only for the extraction of a single mineral; but there might be several minerals present on an actual tract, and the tract might have conflicting uses in farming, aesthetics, or recreation. The multiple mineral problem is complex and will not be discussed here, but the following can be said about conflicting uses of land possessing recreational or aesthetic value. Much extensive mineral extraction occurs in remote areas where recreational and aesthetic attributes are irreversibly damaged by mining. The effect of strip mining in Appalachia or the proposals to mine oil shale in Colorado are examples which indicate the damaging effects on the natural environment. Aesthetic and recreational values are hard to assess, but it is understandable that the value of natural environments rises as their numbers dwindle and man-made wealth accumulates. A house owner with two cars can afford to worry about the environment and can spend a higher fraction of his income on recreation. The market does not respond adequately to the decreasing supply and rising demand for natural environments because the environments are public goods in a broad sense, enjoyed by present and future generations. When comparing the benefits of mineral extraction, which might amount to lower copper prices for fifty years, with the benefits of preservation for future generations, the government should decide to leave some of its mineral properties undeveloped, preferably saving large tracts rather than small. Whenever possible it should locate oil wells and copper pits in one place instead of scattering them around on every other square mile so that they are always in sight. The problems of instituting minerals policy in British Columbia or anywhere else are more complex than the ideal model suggested. [25]

POLICY CONCLUSIONS

The ideal model demonstrates the basic efficiency of the market and advises against encouraging or discouraging mineral exploitation or distorting its pattern in any way. In the absence of specific market imperfections or distortions, the workings of the market should be accepted and mineral tracts freely offered for lease. Any number of leasing arrangements could be used, ranging from outright sale of mineral rights to sale of minerals by the ton. When a degree of realism is considered, a more active role for government is suggested and some policy directions are indicated.

The size and pattern of leases offered to firms become important. Information spillovers and scale economies in the early exploratory work require the leasing of large tracts. Claimstaking, chequerboard leasing, and lottery systems should be avoided because they frustrate the capture of spillovers and prevent the utilization of scale economies. From an environmental standpoint, they also may spread the visual disamenities of mining too widely. It might be better to have all of those ugly derricks and pits in one place rather than spread around the countryside.

The leasing of large tracts may aggravate risk problems, exclude small Canadian firms, and reduce the number of bidders on mineral properties. As Mead has suggested, [26] these difficulties probably arise more often in virgin areas than in known mineral districts, where the outcomes are more certain. Where difficulties do arise, the government should avoid lease arrangements that require bonuses and other large front end payments. Royalties and profit sharing are preferable because they transfer some of the risk to the government, and the government is assured a continuing share in any successful ventures—a political necessity in most cases. The government may want to share even more of the risk, internalize the information spillovers, and help out the smaller firms by performing the early exploration itself before leasing out the tracts. The smaller firms can better participate once the big risky exploratory operations are out of the way. This may advance nationalistic goals for Canada as well as increase the number of bidders to maximize government revenues. Finally, mineral leasing policy must operate in a complicated economic and political environment; no policy can be rejected out of hand because it fails the optimality tests of some idealized model. All things considered, a distortionary policy may be the best one to use.

Appendix: A Solution of the Model

The Crown wants to pick the time path of output for the ith tract that maximizes the present value integral, (1). According to the Pontryagin Maximum Principle,[27] for a maximum it is necessary that the Hamiltonian,

$$\tilde{H}_i e^{rt} = P\dot{X}_i - C_i(\dot{X}_i, X_i, t) - q_i\dot{X}_i , \qquad (14)$$

be maximized at all times. To define q_i, the shadow price of the resource, let $PV_i^*(X_i(t),t)$ be the maximum present value for activities from time t forward or

$$PV_i^*(X_i(t),t) \equiv \underset{\dot{X}_i}{\text{MAX}} \ [\textstyle\int_t^\infty P\dot{X}_i - C_i(\dot{X}_i,X_i,t)\overline{e}^{r(\tau-t)}_{d\tau}] , \qquad (15)$$

where τ is a dummy variable of integration. We can then define q_i as the decrease in the present value of the ith tract caused by the removal of one ton of mineral, or

$$q_i \equiv - \frac{\partial PV_i^*}{\partial X_i} \qquad (16)$$

(q_i is also called the marginal user cost). At any point in time, the Crown picks the rate of output, $\dot{X}_i$, that maximizes the Hamiltonian by setting the partial derivative equal to zero,

$$\frac{\partial \widetilde{H}_i e^{rt}}{\partial \dot{X}_i} = P - \frac{\partial C_i}{\partial \dot{X}_i} - q_i = 0 , \qquad (17)$$

which means setting marginal extraction cost plus marginal user cost equal to the price.

The Maximum Principle also says that the time path of q_i follows the differential equation

$$\frac{d}{dt} q_i = \frac{\partial \widetilde{H}_i e^{rt} + q_i r}{\partial X_i} = - \frac{\partial}{\partial X_i} C_i + q_i r . \qquad (18)$$

Since $\frac{\partial}{\partial X_i} C_i = 0$ and $q_i > 0$ before extraction begins, q_i grows at the rate r during this period. While extraction is occurring, $\frac{\partial}{\partial X_i} C_i > 0$, and q_i grows at a rate less than r and may even shrink. The optimal time paths of $\dot{X}_i$ and q_i vary greatly from example to example, and are not obvious from inspection of the equation. An actual computation follows:

Suppose that a mythical character, Grubstake Pete, has leased a small 100-ton copper deposit in the Highland Valley. The price of copper follows the relationship, $P = \frac{1}{2}t$, and Pete's cost function is $C_i(\dot{X}_i) = 5\dot{X}_i + \frac{1}{2}\dot{X}_i^2$. (Note that t and X_i are not arguments of the function, so costs are not affected by technical progress or depletion.) Discounting at the rate of 10 per cent, Pete wants to maximize the present value integral

$$PV_i (0) = \int_0^\infty [\tfrac{1}{2}t\dot{X}_i - 5\dot{X}_i - \tfrac{1}{2}\dot{X}_i^2] \, \bar{e}^{.1t} dt , \qquad (19)$$

subject to the constraints $\dot{X}_i \geqq 0$ and $X_i \leqq 100$. His Hamiltonian,

$$\widetilde{H}_i \, \dot{e}^{1t} = \frac{1}{2}t\dot{X}_i - 5\dot{X}_i - \frac{1}{2}\dot{X}_i^2 - q_i\dot{X}_i \,, \tag{20}$$

is maximized by setting its partial derivative equal to zero:

$$\frac{\partial \widetilde{H}e^{.1t}}{\partial \dot{X}_i} = \frac{1}{2}t - 5 - \dot{X}_i - q_i = 0 \,. \tag{21}$$

According to equation (18), the time path of the marginal user cost grows at a 10 per cent rate,

$$\frac{dq_i}{dt} = 0.10q_i \,, \tag{22}$$

so marginal profits grow at a 10 per cent rate during production. If Pete's discount rate is 10 per cent, it makes sense for marginal profits to grow at that rate, because he would be indifferent between a dollar now and $e^{.1t}$ dollars t years from now on a marginal unit of production. From (21) and (22), we find the optimal time path of $\dot{X}_i$ to be

$$\dot{X}_i = \frac{1}{2}t - 5 - q_i(0)e^{.1t} \,, \tag{23}$$

where $q_i(0)$ is the initial value of q_i to be determined by the side conditions on the problem. These require that $\dot{X}_i$ equal zero at the beginning and end of the extraction period and that the deposit be exhausted.[28] By some rigorous trial and error methods, Pete finds that $q_i(0) = 0.25$, which means that an extra ton of ore at time zero is worth twenty-five cents to him. His time path of extraction is graphed in Figure 4.

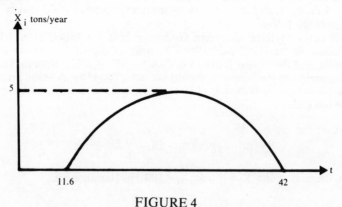

FIGURE 4

Note that he delays mining for a while after marginal profits become positive (at t = 10), because he considers user cost as well as extraction costs on the initial unit of output. He owns the mineral rights and must consider the cost of depleting them.

Notes

1. Harold Hotelling, "The Economics of Exhaustible Resources," *Journal of Political Economy* 39 (April 1931): 140-46, and Frederick M. Peterson, "The Theory of Exhaustible Resources: A Classical Variational Approach," (Ph.D. diss. Princeton University, 1972). For a similar demonstration with a discrete-time model, see Milton C. Weinstein and Richard J. Zeckhauser, "The Optimal Consumption of Depletable Natural Resources," *Quarterly Journal of Economics* 89 (August 1975): 391-92.

2. L.S. Pontryagin et al., *The Mathematical Theory of Optimal Processes* (New York: Interscience Publishers, 1962).

3. Anthony Scott, *Natural Resources: The Economics of Conservation* (Toronto: University of Toronto Press, 1955), pp. 5-8.

4. Sometimes mining starts and stops as prices fluctuate. For the case of lead, see J.M. Heineke, "The Demand for Refined Lead," *Review of Economics and Statistics* 51 (November 1969): 374-78.

5. The maximum present value at time t is given by the infinite integral

$$PV_i^* = \int_t^\infty [P\dot{X}_i^* - C_i] e^{-r(\tau - t)} d\tau ,$$

where τ is a dummy variable of integration. The integral is differentiated with Leibniz's formula. See I.S. Sokolnikoff and R.M. Redheffer, *Mathematics of Physics and Modern Engineering* (New York: McGraw-Hill Book Company, 1958), p. 262.

6. This paper does not consider the important problems that arise when there is more than one discount rate in the economy. For a discussion of the relationship between the rates of return on extractive resources and other assets, see Robert M. Solow. "The Economics of Resources or the Resources of Economics," *American Economic Review* 64 (May 1974): 1-14.

7. If the firm were guaranteed its partnership role for the life of the resource, it would maximize the present value of its receipts. If it received a fraction, b, of the profits, the present value of its share would equal $b\int_0^\infty [P\dot{X}_i - C_i] e^{-rt} dt$, which is just the original present value integral, (1), multiplied by a constant. By referring to the Appendix, we find the Hamiltonian for the firm to be

$$\tilde{H}e^{rt} = b(P\dot{X}_i - C_i) - bq_i\dot{X}_i ,$$

which would be maximized by the same value of $\dot{X}_i$ whether b equaled 0.01 or 1. Therefore, a firm would behave optimally in the ideal model whether it received 1 per cent or 100 per cent of the profits due to the owner of the resource.

8. Gregg Erickson, "Alaska's Petroleum Leasing Policy," *Alaska Review of Business and Economic Conditions* 7 (July 1970): 9-12.

9. Lewis Cecil Gray, "Rent Under the Assumption of Exhaustibility," *Quarterly Journal of Economics* 28 (May 1914): 466-89.

10. Erickson, "Alaska's Petroleum Leasing Policy," pp. 6-8.

11. For a demonstration of the efficiency of market behaviour under conditions of risk neutrality, see Weinstein and Zeckhauser, "The Optimal Consumption," pp. 390-92.

12. As the literature on bidding strategy has shown, firms make errors in estimating the expected value of tracts, and they protect themselves from these errors by bidding less than their expected value estimates. Otherwise they would win a disproportionate share of tracts on which they had overestimated the value, and they would lose money on average. This is quite different from risk aversion. See E.C. Caper, R.V. Cleppard, and W.M. Campbell, "Competitive Bidding in High Risk Situations," *Journal of Petroleum Technology* 23 (June 1971): 641-53, or M.E. Oren and A.C. Williams, "On Competitive Bidding," *Operations Research* 23 (Nov.-Dec. 1975): 1072-79.

13. Maurice Allais, "Method of Appraising Economic Prospects of Mining Exploration over Large Territories: Algerian Sahara Case Study," *Management Science* 3 (July 1957): 285-347.

14. Louis B. Slichter, "Mining Geophysics," *Mining Congress Journal* 45 (May 1959): 38-39.

15. Louis B. Slichter, "Geophysics Applied to Prospecting for Ores," in Alan M. Bateman, ed., *Economic Geology, Fiftieth Anniversary Volume* (Lancaster, Pa.: Economic Geology Publishing Company, 1955), pp. 885-86.

16. Lee E. Preston, *Exploration for Non-Ferrous Metals* (Washington: Resources for the Future, 1960) p. 32.

17. Ronald A. Shearer, "Nationality, Size of Firm, and Exploration for Petroleum in Western Canada, 1946-1954," *Canadian Journal of Economics and Political Science* 30 (May 1964): 211-27.

18. "Smaller Firms Score Big in Huge North Slope Sales," *Oil and Gas Journal* 67 (September 1969): 23-25.

19. Walter J. Mead, *Competition and Oligopsony in the Douglas Fir Lumber Industry* (Berkeley: University of California Press, 1966).

20. Darius Gaskins and Barry Vann, "Joint Buying and the Seller's Return: The Case of OCS Lease Sales," Mimeographed (Berkeley: University of California, 1975).

21. In fact, the relationship between the number of bidders and the size of the winning bid can be explained without assuming collusion: see Robert Wilson, "Price Formation Via Competitive Bidding," (Mimeograph, Stanford University, Palo Alto, 1975), or M.E. Oren and A.C. Williams, "On Competitive Bidding," *Operations Research* 23 (November-December 1975): 1072-79.

22. *Washington Post*, 4 October 1974.

23. The gross depletion allowance allows the producer to deduct a percentage of his revenues from taxable income. The net depletion allowance of the type used in Canada allows the producer to deduct a percentage of net income. For a discussion of minerals tax policy, see Mason Gaffney, ed., *Extractive Resources and Taxation* (Madison: University of Wisconsin Press, 1967). Since this paper was written, depletion allowances have been drastically curtailed in the United States.

24. Let ϕ be the corporate profits tax rate, d x 100 per cent be the percentage allowed

on depletion, and g x 100 per cent be the royalty percentage. The firm maximizes its present value, which is given by the integral

$$PV_i(0) = \int_0^\infty [P\dot{X}_i - C_i - gP\dot{X}_i - \phi(P\dot{X}_i - C_i - gP\dot{X}_i - dP\dot{X}_i)] e^{-rt} dt$$

$$= \int_0^\infty [(1-\phi)(P\dot{X}_i - C_i) + (\phi d + \phi g - g)P\dot{X}_i] e^{-rt} dt$$

Note that the royalty is assumed to be deductible.
If the expression($\phi d + \phi g - g$)vanishes, the integral becomes the usual one (1) multiplied by a positive constant,($1-\phi$).As we learned in Note 7, such a constraint does not alter the behaviour of the firm, which would still behave optimally. The expression vanishes if $d = g$ and $\phi = 0.50$, but there are many other combinations of tax, depletion, and royalty rates that make it vanish and produce optimal behaviour.

25. For a discussion of environmental amenities in general and molybdenum mining in particular, see John V. Krutilla and Anthony C. Fisher, *The Economics of Natural Environments: Studies in the Evaluation of Commodity and Amenity Resources* (Baltimore, Johns Hopkins Press, 1975).

26. See Walter J. Mead's article in this volume, "Cash Bonus Bidding for Mineral Resources."

27. Pontryagin et al., "Mathematical Theory of Optimal Processes" and Kenneth J. Arrow, "Applications of Control Theory to Economic Growth," in Dantzig and Veinott, eds., *Mathematics of the Decision Sciences, Part 2* (Providence, R.I.: American Mathematical Society, 1968), pp. 85-119.

28. Pontryagin et al., Mathematical Theory of Optimal Processes: 257-316.

Cash Bonus Bidding for Mineral Resources

WALTER J. MEAD

Mineral leasing policy alternatives arise out of the fact that governments own mineral resources but, in general, do not engage in mineral resource recovery and processing. Hence, a need arises to transfer publicly owned resources to private enterprise at a price which will reflect the "fair market value" of the resource. The following analysis will, first, explore the problems to be solved by a bidding policy and, second, evaluate the cash bonus method of bidding for mineral leases.

THE PROBLEMS TO BE SOLVED

There are three problems which must be solved by any leasing system. First, the leasing system must as objectively as possible determine who or what firm is to be given the right to exploit publicly owned mineral resources. Second, a price must be determined which the lessee is to pay to the government for the right to recover mineral resures held in trust for its citizens. Third, assuming that a nation wishes to economize on the use of its scarce resources and to maximize the standard of living of its citizens over time, the leasing system must result in an efficient method of production. As a prerequisite to a discussion of mineral leasing alternatives there should be a clear statement of the goal(s) to be achieved. It is probably true that economists as a group have a preference for a single goal, declaring it to be one of economic efficiency. Natural resources available to any economy are scarce by definition. Achieving the highest possible standard of living requires that scarce resources be utilized with a maximum of efficiency. If resources are sold at a price below their true value, then the products into which they are converted may also be underpriced. If demand elasticities are less than zero, then the flow of resources into products and the flow of products within the current period will be excessive. Present overconsumption of products and resources will be at the expense of future consumption.

One way of achieving maximum economic efficiency is to price all resources at their "fair market value." Such pricing allows a government the opportunity of capturing the economic rent. Resources should be sold for the difference between future revenues and costs, appropriately discounted to their present value. The economic principle relating prices, costs, and money flows at different points in time in order to estimate present value (*PV*) is shown in the following formulation:

$$PV = \sum_{i=0}^{n} \frac{P_i Q_i - C_i}{(1+r)^i}$$

$P_i Q_i$ is the value of the gross income flow at different points in time, C_i represents associated costs, and r represents the interest rate at which future money flows are discounted to the present. The formula clearly shows that higher future prices will increase present values while higher future costs will lower present values. Further, the greater the uncertainty and risk associated with production, the smaller will be present value. Firms utilize some variation of this present value formula in calculating their individual bids. Estimates of the quantity of minerals recoverable from a given tract will, of course, vary widely from firm to firm.

If mineral leases are sold for less than the fair market value as indicated in the above formula, then resources may be used at an excessively rapid rate, and the public, as owners of the resource, will fail to receive their full economic rent. On the other hand, if mineral resources are sold at prices in excess of the fair market value then, in the long run, some operators will be forced out of business. Use of such mineral resources in the present period will be at a suboptimal rate and the public owners will receive more than their normal economic rents.

In the past, Canada apparently has transferred some of its mineral and timber resources through various negotiated transactions rather than by utilizing the auction market approach. Similarly, other foreign governments have traditionally taken the negotiated sale approach in entering into long term oil concessions.

There are major problems involved in the negotiated approach. The correct present value of natural resources is extremely difficult to ascertain. There is no objective test in advance of ultimate production that can indicate the precise present value of mineral resources. By their nature they must first be discovered. Their presence, quantity, and quality are in doubt. With the government as the seller, negotiating with a single buyer, traditional problems of bilateral monopoly are encountered. The seller is interested in maximizing price, while the buyer is interested in minimizing price. Given this uncertainty plus opposing objectives, the civil servant is placed in a difficult position.

A visitor to Canada is reluctant to criticize Canadian experience which has circumvented the market place. Fortunately there is abundant experience within the United States to indicate the shortcomings of the negotiated approach to pricing. We may formulate two general laws which seem to govern when prices are determined or may be influenced by administrative judgement. First, the buyer will always complain. If the buyer believes that market prices can be reduced by protesting that they are too high, then complaints based on the argument that the operator cannot make a "fair profit" because prices are set too high could be endless. In the timber context, there are two cases where elaborate reports have been written protesting the high price of timber. One, presented by the Simpson Timber Company, protested against the high price of stumpage set by the United States forest service for the Shelton Sustained Yield Unit Agreement. Timber, in this case, is not sold at auction; its price is determined by the United States forest service. In the second case, the Edward Hines Lumber Company protested against the high cost of timber for its southeastern Oregon lumber mill. The timber was sold at an auction where competition was so weak that, in effect, it was sold at the administratively determined minimum price. By protesting, the company apparently felt that minimum prices could be reduced. In this instance, local community help was solicited on the grounds that if the company failed to make a fair profit, it would be forced to curtail operations. Under auction market procedures the government is relatively free from constant complaint and protest, because it is the impersonal market that determines the price rather than a civil servant. Under auction bidding procedures the buyers themselves set the price in competition with one another.

A second general law is that, where prices are set through administration, the government will always set prices short of the fair market value. A bureaucracy will rarely choose the path that makes its position unpleasant. Low prices are believed to generate less criticism and complaint than high prices. Where there is no auction market to test administrative judgement concerning the fair market value, we have no means to prove the second law. Sales of timber in the United States offer an opportunity to test the administratively determined price. Timber is sold by the forest service on the basis of an appraised fair market value, which becomes the minimum price acceptable to the government. Auction bidding begins at this price. In the four years from 1959 through 1962, competitive bidding for timber in the United States Douglas fir region produced an average high bid price that exceeded the forest service statement of fair market value (the appraised price) by 46 per cent.[1] In this case, the interests of the public were protected, at least in part, by reasonably effective competition. In the absence of this competitive check it is quite likely that the appraised prices would have been even lower. The shortcomings of the negotiated approach should lead to auction bidding wherever competition is possible.

CASH BONUS BIDDING

Before bidding can take place, a decision must be made between oral and sealed bidding. Bidding in either form may start with a stated or unstated minimum acceptable price. In the case of timber sales in the United States, the minimum acceptable price is given by the appraised price, and most timber auctions are conducted under oral auction procedures. On the other hand, in the case of oil and gas leases conducted by the federal government in the United States, the minimum acceptable price is not published, and bidding is normally by sealed bidding procedures. The government retains the right to "reject any and all bids." After bids have been received, it determines whether or not the high bid was adequate.

The factors important in choosing between oral and sealed bidding methods are as follows:

a. Of prime importance is the extent of competition. If competition is weak, then sealed bidding with its element of uncertainty makes collusive arrangements more difficult to enforce. Under sealed bidding rules there is no second chance to bid at any given sale. In contrast, under oral bidding procedures, a collusive arrangement can be policed by the participants during bidding. Further, there is always doubt about how many bidders may appear at a given sale. In oral bidding where only one bidder is present, he will bid the minimum; whereas, in sealed bidding a bidder would probably offer an amount which he believes will win the sale under conditions of more than one bidder.
b. In the timber industry where fixed investments in milling facilities normally exist prior to sales, the buyer needs a means of ensuring access to specific raw materials and specific locations. Oral auction procedures provide this means through the opportunity to cast reaction bids. In contrast, in oil and gas bidding fixed investments are made after winning a sale, hence there is less need to protect one's position through the opportunity to react to the bids of others.
c. Where the severed resource is relatively immobile, as in the case of timber, it is of greater importance that a specific nearby sale be obtained; therefore the oral auction procedures are more appropriate. In the case of oil and gas, the severed resource is highly mobile, so obtaining a specific sale is of less importance. In this case sealed bidding is not disadvantageous.
d. Where the resource to be auctioned is not homogeneous, it may be necessary for a firm to obtain a specific sale. Where this is true, the opportunity to make more than one bid to protect one's need for a specific type of resource may be of great importance. Only oral bidding facilitates this subsequent bidding opportunity.
e. Financial planning often requires that a firm carefully limit its financial exposure. Where this is necessary, oral bidding offers greater control over a total resource financial commitment. In the case of sealed bidding,

firms may be unexpectedly successful and in the process win more sales
than were desired or can be successfully financed. On the other hand, a
firm's sealed bidding may be totally unsuccessful so that it becomes
undercommitted. This shortcoming of sealed bidding may be corrected
where resources may be freely transferred among interested buyers. This
procedure is normally followed in the case of oil and gas leasing in the
United States.

f. Oral bidding requires more on-the-spot decision making than does sealed
bidding; therefore, oral bidding requires that a higher level of executive
talent be present at the moment of the auction. In contrast, decisions
made on the basis of a sealed bid offer no opportunity for subsequent
action on the auction floor; therefore, the presence of expensive executive
talent is not necessary.

g. The "free rider" is a problem for serious bidders under oral bidding
conditions. A serious bidder will carefully examine the potential
productivity of a proposed lease sale. This may, as in the case of
minerals, require large investments. Under oral bidding conditions, a
"free rider" can observe who is bidding, then, if he is confident that they
have done their homework, he can continue to outbid them until they
reach their maximum and he will win the sale. His purchase is therefore
based on someone else's calculations and he, in turn, has saved the cost
of the pre-exploration appraisal. Sealed bidding does not offer the free
rider the same opportunity.

Once a decision has been made in favour of oral or sealed bidding, then a
choice must be made on the object of bidding. A cash bonus bid is one
alternative. Additional alternative bidding objects are shown in Table 1.[2]

Bonus bidding is the standard procedure used by the United States
government in all of its Outer Continental Shelf (OCS) programmes. Using
the present value formula given earlier, potential bidders presumably
estimate the present value of the probable mineral recoverable from a tract
of land. The formula provides for adequate recovery of capital and
compensation for risk, uncertainty, and profit.

One strong advantage that can be claimed on behalf of a bonus system
relative to royalty bidding is that it requires a lump sum payment and
correspondingly modest royalty payments. Because royalty payments are
due on each barrel of oil or unit of natural gas produced (or other
mineral), such charges become part of the marginal cost. At the margin of
production this is a transfer cost rather than a real social cost. Royalty
bidding thus leads to premature abandonment of an oil or gas well. To the
extent that royalty payments are required in addition to the cash bonus,
there will be premature abandonment of the lease.

The disadvantages of bonus bidding are numerous. First, while the technology for oil exploration prior to drilling has been advanced in the last century, exploration is still subject to extremely high risk. Drilling is the only definitive test to determine the presence of oil or gas. Thus, bonus bids must be submitted by bidders and accepted or rejected by the government when neither the buyer nor the seller knows whether and in what quantities oil is present. This places the seller in a position of accepting millions of dollars for nothing but the right to spend several more millions drilling potentially dry holes. In cases in which a rich oil field is found, returns to the lessee will be and must be very high.

Second, under current procedures a bonus must be paid when the bid is submitted. When the bonus bid is large, it will represent a very heavy cash drain to the bidder far in advance of any revenue which may be generated from the oil or gas produced from the lease. This significant *front-end loading* of capital costs effectively excludes a small operator from winning leases as a solo bidder, creating an additional barrier to entry into the oil and gas production market. To overcome this entry barrier, firms commonly form joint ventures and bid jointly for a lease.

Third, because the bonus is calculated on a present value basis, the government is forced to accept discount rates used by private enterprise. If private enterprise discount rates are unreasonably high from a social standpoint, then bonus payments to the government will be correspondingly low.

Possible variations of the bonus bidding form are shown in Table 1. The present United States system includes fixed royalty requirements (typically 12½ per cent or 16⅔ per cent of wellhead value). However, a bonus bid might be paired with a sliding scale royalty requirement permitting the royalty rate to be reduced as a field declines in productivity. As the point of economic abandonment is approached, the royalty rate might be reduced substantially or even eliminated. This procedure would, in turn, eliminate a marginal cost of production that is not a real social cost, and it would permit continued production from a field until the real marginal costs equaled the marginal value of production. This is the optimum point for well abandonment from an economic point of view. If at the time that a bonus bid was submitted all bidders understood that the royalty rate would be reduced to zero under the conditions specified above, the present value of the lease would be increased by an amount equal to the present value of reduced future royalty payments. Thus a tradeoff would be effected from royalty payments to bonus payments. The principal impediment to a sliding scale lies in the difficulty of clearly identifying various points at which royalty rates would be reduced. The lessee would have an economic incentive to manage his production in such a way that minimization of royalty payments would be an operating objective, rather than economic efficiency.

TABLE 1

ALTERNATIVE BIDDING FORMS

Bonus Bidding
  a. with a fixed royalty requirement
  b. with a sliding scale royalty requirement
  c. without a royalty
  d. with or without a rental payment
  e. with a profit share
  f. with delayed bonus payments

Royalty Bidding
  a. flat (nonvariable) royalty
  b. sliding scale royalty
  c. with a fixed bonus requirement or no bonus

Profit Share Bidding
  a. net profit or gross profit
  b. with fixed bonus requirement or no bonus
  c. with a royalty requirement or no royalty

Combination of Bonus and Royalty Bidding

Bidding on the Work Programme

The royalty problem, together with the administrative problem of reducing royalty rates under a sliding scale, might be avoided entirely by using a bonus bid without a royalty payment. However, this procedure would simply magnify all three of the problems associated with bonus bidding listed above.

Present procedures in the United States include modest rental charges payable between the points of sale and production. When production begins, rental payments cease and royalty payments take over. Rental payments in OCS oil and gas lease income are insignificant. In 1972, they amounted to 0.3 per cent of total revenue from such leases.[3] The rental requirement apparently was introduced to motivate the lessees toward early production. If they were of significant size, this result would in fact occur, because rents cease when production begins.

To overcome the front-end-loading problem, provision might be made for delayed payment of the bonus. The problem that would follow from this procedure is that in some cases where no minerals were found, lessees would elect a bankruptcy route. In this event, an unfair bidding situation would be created. Responsible firms in business on a perpetual basis would not follow a bankruptcy procedure and would, therefore, be at a bidding disadvantage with respect to others that contemplated bankruptcy in the event of a "dry hole."

A bill currently pending before the United States Congress provides for a 55 per cent fixed share of net profits in lieu of the existing fixed royalty payment accompanying the bonus bid. The winner would still be determined on the basis of a cash bonus. A profit share payment would avoid the above problems associated with royalty payments. As a given lease approaches exhaustion and its point of economic abandonment, profits would also approach zero and payments would decline proportionately to zero. If the profit share was calculated on the basis of net accounting profits including fixed costs, then the profit share payment would decline to zero prior to the point of economic abandonment. The latter point is reached only when marginal cost (not total costs) equals the marginal revenue. There is nothing wrong with this system providing both parties understand how it works and bidders understand it at the time they submit their bonus bids. The proposed 55 per cent profit share is high and is likely to lead to inefficient operations. A profit share payment is approximately the same as an income tax on each well and is additional to the existing income tax. When the profit share payment is added to the existing income tax, a large part of the penalty for wasteful operations will have been shifted from the operator to the government. While a bonus bid paired with a fixed profit share payment has merit, a 55 per cent profit share added to normal income taxes is inappropriately high from an economic point of view.[4]

Some data are available to permit a partial evaluation of the effectiveness of bonus bidding with a fixed royalty. The United States experience with OCS bidding provides a record of thirty-five oil and gas lease sales during the period November 1954 to 29 May 1974. In addition, three sulphur lease sales and two salt lease sales have been conducted on the OCS. The record may be evaluated in terms of the number of bidders competing for each sale, the conditions of entry of new firms, the record of joint bidding, the extent of concentration among winning firms, the trend in price bid per acre, the resale record of tracts where the initial bid was refused by the seller, and the rate of return earned by the winning bidders. Data pertaining to OCS bidding as follows:

a. For oil and gas lease sales there has been an average of 3.6 bidders competing for each tract receiving bids. The trend from 1954 to date has been one of increasing bidder activity. From 1954 to 1966 the average number of bidders per tract was 2.7. From 1967 to date the average increased to 3.9. From the seller's point of view, even more bidders would be preferred. Given the fact of relatively few bidders, sealed bidding procedures would appear to be more appropriate than oral auction.

b. Entry into the oil and gas auction markets appears to be relatively free. In the first 1954 federal lease offshore from Louisiana, 199 tracts were offered. Ninety-seven of these tracts received 327 bids from 22 different firms, some of which bid in joint bidding combinations. From 1954 to 28 March 1974, an additional 110 firms won tracts as solo bidders or joint bidders with 1 or more other firms. Thus, in addition to the unsuccessful bidders who also perform a competitive function in the bidding process, there were 132 separate firms participating as winning bidders in thirty-three OCS lease sales.[5]

c. Entry by relatively small firms into OCS lease sale bidding is facilitated through joint bidding. Joint bidding by two or more firms each unable to bid solo has the effect of increasing competition. On the other hand, when two or more large firms fully able to bid separately combine to submit a single bid, the effect may be to reduce the number of competitors. However, if through joint bidding, even among large firms, a combination of, say, four firms bids more than four times as frequently as the individuals would have bid solo, then the effect of joint bidding can again be procompetitive.

d. The record shows some tendency toward concentrating winning OCS bids in relatively few hands; however, the extent of concentration also appears to be declining over time. For the nineteen oil and gas OCS lease sales which took place from 1954 through 1966, the eight largest buyers, sale by sale, purchased 85.5 per cent of the tracts. In the fourteen sales from 1967 to 28 March 1974, the percentage of total tracts purchased by the eight largest buyers declined to 62.0.[6]

Using the 184 leases issued in the 1954 and 1955 Louisiana oil and gas lease sales, a multiple regression analysis tested the proposition that firm size was positively related to the high bid by tract as the dependent variable. If large firms are able to outbid smaller firms, then one would expect a positive relationship. The regression analysis revealed no significant relationship between size class of firm (the eight big firms versus all others) and the amount of the winning bid. The same regression equation revealed that the high bid was also independent of whether firms bid jointly or solo. Further, the most significant independent variable related to high bid was number of bidders; the greater the number of bidders competing for any given tract, the higher will be the resulting winning bid. The total value of oil and gas production accumulated through 1967 was also positively related to the high bid. As one would expect the number of acres in the tract leased is also related to the high bid. Estimated water depth as a proxy for development cost was not significantly related to the high bid.[7]

e. Data on the average price bid per acre indicates that with the passage of time the effective high bid per acre has increased substantially. For the entire period 1954 through to 28 March 1974, the average high bid per acre amounted to $1,257.50. For the 1954-1966 period the average was

$301.71 per acre. This increased more than sevenfold to $2,219.90 per acre for the period beginning in 1967. This increase is only partially accounted for by higher crude oil prices. The average price of crude oil increased from $2.89/bbl. in the earlier period, to $3.69/bbl. in the later period. Even this increase would be offset by an unknown decrease in the probability of finding oil, and by increased costs of exploration and production.

f. Lease sales through 1 October 1964 show that of the 1,377 tracts receiving bids, seventy-eight high bid offers were rejected by the government. Subsequently, 26 of these tracts were reoffered and leases awarded. For these 26 tracts, the initial rejected high bid average amounted to $42.41 per acre. The subsequently accepted high bid on resale averaged $411.38. Thus, where bids were found to be inadequate and subsequently reoffered, competition increased bonus payments on these rejected tracts nearly tenfold.

g. The most conclusive test of the workability of cash bonus bidding based on the United States record of OCS oil and gas lease sales is in terms of the rate of return on capital earned by the successful bidders. An analysis has been made of 184 offshore Louisiana oil and gas tracts leased in 1954 and 1955. Precise data are available on bonus payments, rental payments, oil and gas royalty payments, and production of oil and gas during the period from 1954 through 1967. Cost estimates were made for exploration, well drilling and equipment, and operation. Annual cost and annual wellhead values were discounted to obtain a net internal rate of return. The calculations indicate that these early OCS leases generated a 7.5 per cent before tax rate of return to the lessees.[8] Given the fact that oil companies pay relatively low U.S. income tax rates, the after tax rate of return would be only modestly lower than the 7.5 per cent before tax rate of return. This net yield clearly does not reflect monopoly power; it shows an excessive degree of competition.

On the basis of this evidence we conclude that competitive bidding for oil and gas leases is sufficiently strong to protect the public interest in obtaining competitive values for its oil and gas resources. This conclusion is further supported by evidence presented above indicating an increase in the average number of bidders and a substantial increase in the average price bid per acre for oil and gas leases.

CONCLUSIONS

This article has examined the problems to be solved by any leasing system used to transfer publicly owned mineral resources to private firms for processing. The cash bonus bidding system has been used extensively in the United States, particularly in the leasing of OCS mineral resources. That record has been examined in some detail. While cash bonus bidding embodies problems which have been identified, it also appears to be an economically efficient method of resource conveyance. The United States record indicates that competition has been effective, if not overly effective, in permitting the government to capture the full economic rent. In addition, bonus bidding avoids a major problem of a popular alternative, that of royalty bidding. It appears to be far superior to a negotiated approach in solving the three critical problems of resource leasing: selecting the operator, determining a fair market value, and creating a climate for efficient mineral resource recovery.

Notes

1. W. J. Mead and T.E. Hamilton, *Competition for Federal Timber in the Pacific Northwest—An Analysis of Forest Service and Bureau of Land Management Timber Sales* (U.S.D.A., Forest Service Research Paper PNW-64, 1968), p. 4.

2. For a more thorough discussion of the economic issues involved in oral auctions and sealed bidding, see W.J. Mead, "Natural Resource Disposal Policy-Oral Auctions versus Sealed Bids," *Natural Resources Journal* 7 (April 1967): 194-224.

3. U.S. Department of the Interior, Geological Survey, *Outer Continental Shelf Statistics* (June 1973), p. 43.

4. For a more thorough discussion of this point, see W.J. Mead, Testimony Presented before the United States Senate, Committee on Interior and Insular Affairs, Hearings 7 May 1974.

5. The data presented above from the OCS bidding record are from Susan M. Wilcox, "Entry and Joint Venture Bidding in the Offshore Petroleum Industry," (Ph.D. diss., University of California, Santa Barbara, 1975), p. 66.

6. Ibid.

7. The multiple regression equation is as follows:

$$Y = -9.5809 - 0.2279X_1 + 0.0229X_2 + 0.1383X_3 + 0.1235X_4$$
$$(0.1513) \quad (0.0111) \quad (0.1701) \quad (0.0544)$$
$$+ 0.408X_5 + 0.0357X_6$$
$$(0.0253) \quad (0.0235)$$

where Y is the high bid and the unit of measure is $100,000, X_j is the size class of the high bidder coded as 10 for instances where the high bidder is one of the big firms and as zero for all other firms, X_2 is the total value of all oil and gas production accumulated up to the end of 1967 and the unit of measure is $100,000, X_3 is the corporate structure of the high bidder coded as 10 for a joint venture and zero for a single firms, X_4 is the number of acres with a unit of measure in 100 acres, X_5 is the number of bidders per sale multiplied by 10, and X_6 is the estimated water depth. This equation accounts for 62 per cent of the total variability in the high bonus bid. The standard error of estimate is shown in parentheses: see Nossaman-Waters, *Study of the Outer Continental Shelf Lands of the United States*, vol. 1(1968), p. 553.

8. Ibid., p. 56.

Comment

HAYNE E. LELAND

Walter Mead presents a well-argued case for preferring cash bonus bidding to alternative methods of selecting a lessee. In these comments, his analysis is extended to consider explicitly an important element of the mineral leasing environment: the presence of uncertainty. When the effects of risk and risk aversion are analyzed, one can better assess the relative merits of cash bonus bidding.

No one denies the existence of uncertainties associated with natural resource exploration, development, and production decisions. The presence of uncertainty is also important in that changes in risk alone will lead to substantial changes in a firm's behaviour. It is uncertainty which has led the typical transfer agreement for natural resource rights—the lease contract— to differ markedly from the typical transfer agreement for most other assets —the outright sale. In contrast with outright sales, leases normally contain "conditional payments" such as royalties or profit shares. If there were no uncertainties about the value of a lease tract, there would be no need to bother with royalties, profit shares, or other forms of payment beyond the initial cash price. As shown below, "conditional payments" are a response to the detrimental effects of risk on the decisions of private firms.

Risk affects the decisions of firms in several ways. Making the well-supported assumption that firms tend to be risk averse, greater uncertainty can be shown to have the following effects:

a. Firms will tend to explore less. Exploration can be viewed as an investment with a risky return. As the returns become more risky, investment will decrease. In the long run, less exploration means less discovery of resources.

b. Firms will extract the discovered resources too rapidly. Leaving a resource in the ground subjects it to future price uncertainty. Even if future prices are expected to rise, risk aversion may call for greater current production.

c. Firms will bid less for leases. Just as exploration is a risky investment, so is a lease bonus payment. Greater uncertainty will lead to lower bids, even if expected return is unaltered.

These responses to a random environment are detrimental from society's viewpoint.[1] Being better able to bear risk, the government would like to minimize the undesirable effects of uncertainty and risk aversion on firms' decisions.

The government could undertake several courses of action to reduce the welfare losses associated with risk averse behavior by firms. Many of these options would necessitate greater government control or outright government ownership. But short of such controls, the government still has means of reducing risks to firms. It can reduce uncertainty in at least three ways:

a. by means of a tax structure with high rates when circumstances are favourable to the firm, and low rates when unfavourable;

b. by making more information available to firms when critical decisions must be made (for example, through government sponsored seismic studies, or through "chequerboard" leasing schemes which make information on initial tracts available to second-round bidders);

c. by risk sharing through the terms of the lease contract.

While options (a) and (b) merit further examination, the focus here is on (c), risk sharing through the terms of the lease contract.

Most lease contracts contain unconditional and conditional payment clauses. Unconditional payments such as lease bonuses do not depend on subsequent events, such as the discovery of the extent of the resource on the tract. If there are no further conditional payments, the firm which wins the bid bears the entire uncertainty regarding the amount of resources, selling price, and cost of production. Thus, the outright sale (or lease with only unconditional payments) transfers none of the risk from the buyer to the seller. Conditional payments are dependent on conditions which are unknown at the time the lease is sold. Royalties, for example, depend on future production (if any) and on the market price at the time of production. Properly chosen conditional payments reduce uncertainty to the firm, since they involve large amounts only in conditions favourable to the firm, and small amounts otherwise.

As the level of conditional payments—which share risk—are raised, the lease bonus bids—which do not share risk—will fall. The result is that the government bears more risk. Viewed in a different perspective, greater weight on conditional payments is equivalent to the government providing partial insurance.

Greater governmental risk sharing will reduce the social welfare loss resulting from risk aversion. Firms will make more economically efficient exploration, development, and production decisions. And the government will enjoy greater expected revenues.

Following our argument to the limit, one might conclude "the more the royalties, the merrier." This overlooks an important problem which Mead makes clear: royalties may have distortionary effects on some firms' decisions. For example, the premature shutdown problem associated with high royalties may more than offset their risk sharing advantages and result in less rather than more social welfare. [2]

One leasing proposal does stand out, however, as having extremely desirable features. This is the profit sharing lease contract. Profit sharing does transfer risk from a firm to the government, and furthermore, if the profit base is properly defined, it will not distort exploration, development, or production decisions. [3] Mead argues that the proposed 55 per cent profit share, when coupled with corporate income taxes, is "too high." I think this dismissal is premature. If the "profit base" includes all relevant costs (including decision costs) as well as revenues associated with the lease, there should be no "disincentive effects" as long as firms truly maximize profits. There is a danger in profit sharing, of course, if the profit base is incorrectly defined or ambiguous. Whether these problems can be overcome is clearly worth future study—a 1 per cent improvement in leasing revenues resulting from better policies would finance several thousand man-years of economic research!

To summarize, the inclusion of uncertainty into the analysis of leasing policies points out the advantages of "conditional payment" clauses. Royalties are the most common form of these payments. But profit sharing clauses, which also lead to risk sharing, are less likely to cause undesirable distortions in a firm's behaviour. Cash bonus bidding should play an important role in determining the lessee, [4] but, as uncertainties increase, unconditional payments should be reduced relative to conditional payments such as profit shares.

A minor criticism involves Mead's interpretation of some empirical results. Finding that the internal rate of return on 184 leases in the Gulf of Mexico (1954-55) averaged only 7.5 per cent, he concludes that "the United States record indicates that competition has been effective, if not overly effective, in permitting the government to capture the full economic rent." Does his sample size warrant such confidence? Given a log-normal distribution of field sizes, most fields will show subequilibrium rates of return (to balance the occasional immense return). If Mead had examined

rates of return to the North Slope leasing during 1964-67, an area about the
same size as the Gulf of Mexico region his sample covered, he would have
reached a much different conclusion. The point is not that the North Slope
is typical, but rather that even a seemingly large geographical area may not
be a sufficiently large enough sample to warrant strong conclusions about
rates of return to oil and gas leasing.

Notes

1. It has been argued that society should be risk neutral with respect to decisions which
generate a relatively small part of its revenue [see K.J. Arrow and R.C. Lind, "Uncertainty and the Evaluation of Public Investments," *American Economic Review* 60 (1970): 364-
78]. Note that if the lessor is such that leasing forms a substantial fraction of its revenue
(for example, the state of Alaska), it may not be optimal to transfer risk from firms to the
lessor. Thus, lease contracts which might be appropriate to the federal government may
not be appropriate for regional governments.

2. Note the parallel with the *moral hazard* problem encountered in the insurance literature.
Moral hazard is a situation in which the presence of insurance (complete fire coverage, for
example) may alter behaviour in an undesirable way (for example, failure to take normal
precautions against fire).

3. For a formal proof of this and other propositions, see H. Leland and Richard Norgaard,
Alternative Leasing Policies for Oil and Gas on the OCS" (Report prepared for the Office of
Energy R and D Policy, National Science Foundation, September 1974).

4. Competitive bidding theory makes clear that undesirable consequences may follow from
royalty or profit share bidding. If there are no bonus payments required, speculative
bidding may lead to extremely high royalties or profit shares being bid, with development
occurring only in the most favourable circumstances. This happens because firms have
little or nothing to lose by bidding high and then failing to explore or develop. Also, profit
share bidding may not ensure that the most efficient firm will win the lease.

Work Commitment Bidding

GREGG K. ERICKSON

One result of the growing concern in the United States over energy matters has been an increased attention to public policies governing the development of Outer Continental Shelf (OCS) oil and gas resources. The institutional structure under which all such development has thus far taken place was established in 1953 by the Outer Continental Shelf Lands Act.[1] This unamended statute provides the Secretary of the Interior with authority to sell oil and gas leases to the public on the basis of cash or royalty bids offered at sealed bid auctions.

The practice of the United States government since the first such sale in 1954 has been to offer relatively small quantities of offshore acreage on an irregular basis, soliciting always cash rather than royalty rate bids. In recent years, the rate at which acreage has moved to market has been accelerating. However, the average per acre bonus received by the government has also increased, partially reflecting worldwide supply conditions. The fact that bids are received in sealed envelopes has resulted in the winning bid being two, three, or several times the amount of the next highest bid.

Among criticisms of present policy is the assertion that this method of lease allocation diverts undesirably large amounts of *front-end money* into the coffers of the government landowner, money that could, would, and should otherwise be used for development of the resource itself.[2] One possible remedy would involve implementation of the existing statutory authority to substitute royalty rate bids, with fixed and presumably low cash bonuses. The problems created by royalty bidding, principally the premature shutdown effect and the potential for speculator induced misallocation of leases, have been well discussed in the literature. More importantly, they are well understood by persons influencing both public and private mineral resource management policies.[3]

An alternative proposed remedy to this same perceived problem is less well understood. Based in part on the method of lease allocation used in the offshore areas of the United Kingdom, it would allocate exploitation rights to the firm that would commit itself to spending the greatest sum in developing the resource. Sealed bids would be solicited as under the present system, but instead of cash the bid variable would be the *work commitment*. Proponents of this system claim that it will divert money the government landowner would otherwise receive via bonuses into exploration and development expenditures.[4] These additional increments of expenditure, it is further suggested, will increase future production to such an extent that the government landowner will be able to recoup the foregone bonus income in the form of the consequentially increased royalty and tax revenue. Ancillary benefits in the form of employment, resource self-sufficiency, and improved trade balances are also sometimes claimed or alluded to.

To an economist these arguments may not seem too persuasive. Nevertheless, no one appears to have devoted much effort to analyzing the economic implications of such a system, and certainly not in a form that would be comprehensive to the noneconomist policy maker.[5] This is unfortunate not only because of the substantial public and private interests involved; the system has significant implications for minerals other than petroleum and in places other than the United States OCS. The purpose here is to provide such an analysis.

EVALUATION OF WORK COMMITMENT BIDDING

In evaluating something new the first step is usually to establish a standard against which it can be measured. In this context, the system of competitive cash bidding has long attracted economists concerned with the problem of natural resource allocation, not only as an ideal against which the performance of other systems might be measured, but as a practical and proven technique for bringing resources into productive employment.

Under an idealized competitive cash bidding arrangement, bidders determine the amount they can afford to offer for a mineral lease by a very simple process: they subtract their expected costs of extraction from their expected revenues. The resulting residual is the maximum the prospective bidder can offer for the tract without buying himself an expected loss. Competition, of course, implies that multiple firms will be preparing bids on each tract.

Assuming no uncertainty about the amount of oil to be found or the price that oil will eventually bring, and disregarding the time value of money, the firm with the lowest expected costs of extraction will be capable of submitting the highest, and thus the winning, bid. This is good from society's standpoint, since it means that the resource will be developed with the minimum expenditure of scarce goods and services. The resource's contribution to economic welfare will be greater than it would have been had the tract been awarded to any of the other, less efficient bidders.

Under a work commitment system each prospective bidder will be asking himself: What is the maximum amount I can promise to spend on the development of this tract and still expect to break even? Since any cash bonus that would have been offered to acquire a tract under the traditional system is no longer necessary, the amount of that bonus may clearly be diverted to the work commitment without raising costs beyond the breakeven point. What is not quite so obvious, however, is that the amount a bidder will promise to spend under the commitment system will exceed the sum of the cash bonus and the amount that he would have allocated to development of the tract under the cash bonus system.

This follows from the fact that any additional increment of expenditure can almost always be spent in a way that will bring about some increase in output from the tract and a corresponding increase in revenue.

This is most easily demonstrated with a numerical example. Let us assume that a prospective bidder, in determining how much of a cash bonus he can offer for an oil and gas lease on a hypothetical tract A, has calculated the relationship between expenditures on development of the tract and expected production, and that the results of his calculations appear as plotted in Figure 1.

The vertical axis in this graph (and those that follow) measures dollars expended in the tract's development, dollars that we assume will be spent for construction of a platform and the drilling from it of wells. The horizontal scale measures the output that results from that expenditure, denominated in millions of barrels of oil. The relationship between those barrels of output and the revenue they bring their producer (at an assumed price of $10 per barrel) is shown, through appropriate choice of scales, by the dashed 45° straight line running upward to the right. By this means, the vertical scale can be used to show the value of output as well as the cost of production.

In Figure 1, the point closest to the origin indicates that with one platform and one well this operator would expend $6 million (vertical scale) producing an output of 550,000 barrels of oil, worth $5.5 million (determined by the intersection of the 45° line with a line drawn vertically from .55 million barrels). Moving upward and to the right, each subsequent point reflects seriatum the increases in expenditures and output resulting from the drilling of additional wells.

The general shape of the curve defined by these points is characteristic of situations where one major input to the productive process (in this case land) is held constant, while other inputs (in this case wells) are varied. The output curve originates at the lower left hand corner, but it rises vertically at first because the initial input of investment is unproductive: a platform and oil well costs a certain amount, and an expenditure of anything less than that threshold amount produces no oil. The cost of subsequent wells is assumed to be $1 million, no matter how many wells are drilled, creating a curve that looks like a staircase where each increment of cost (representing a new well) creates a new step. The fact that the staircase steepens as we move

64

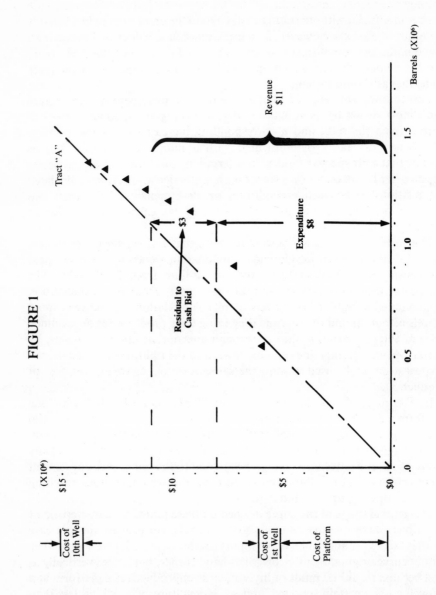

FIGURE 1

to the right is a reflection of the diminishing returns, in terms of oil produced, to each additional well drilled into the fixed geographic area encompassed by the lease.

Naturally the prospective bidder will be looking for the point on this output curve that puts his costs as far below the 45° line (his output-revenue function) as possible. As shown in Figure 1, the maximum cash bid this operator could afford to make on tract A (and still expect to break even) is $3 million, which—if he is the winner—would require him to drill three wells.

Consider now the situation this bidder would face were a work commitment bidding system adopted. The question that now confronts him is: How much can I spend (or how many wells can I drill) on tract A and still break even? The answer is clearly $14 million (representing nine wells), indicated on the right side of Figure 2 by the output curve for tract A.

If the bidder wins tract A under a work commitment system, his oil output will be 1.4 million barrels (Figure 2) as compared to the 1.1 million barrels (Figure 1) that he would have produced had he won the tract in a cash bonus sale.

If the success of a mineral resource management policy is measured by the physical quantities of the mineral educed from the earth, the work commitment bidding is clearly superior. A resource's contribution to economic welfare, however, is not its total output (whether measured in dollars or physical quantities) but is the residual left over when the costs of all inputs to the productive process (other than the resource itself) are subtracted from the value of the outputs. In the case of tract A this residual is maximized at $3 million, when the value of the inputs is $8 million. As the input expenditure is increased above this optimum point, the residual—the resource's potential contribution to economic welfare—is gradually dissipated until, at the point where the value of inputs reaches $14 million, there is no more residual left to be dissipated.

In this particular example, the increase in output that would result from a switch to work commitment bidding ($3 million) happens to equal the amount of the residual. This coincides with the fact that the expenditure of each additional $1 million above $8 million (three wells) contributes exactly $500,000 to revenue. If the incremental contribution of the fourth and succeeding wells were greater, for example $750,000, the slope of the output curve traced by these points would be flatter, as shown by the squares in Figure 3, and the increase in production from a switch to commitment bidding would be much greater. To put it another way, it would take twice as large an increase in expenditure to dissipate the $3 million residual.

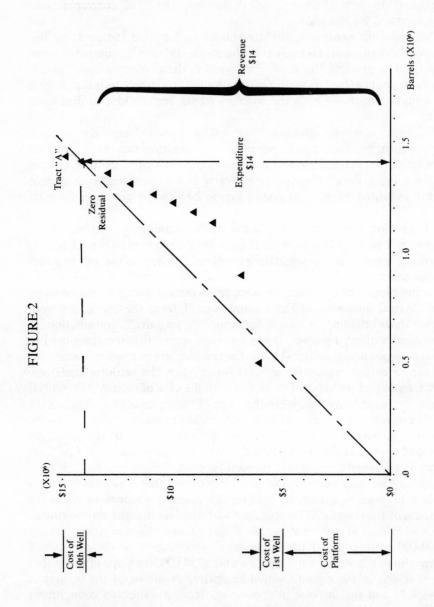

FIGURE 2

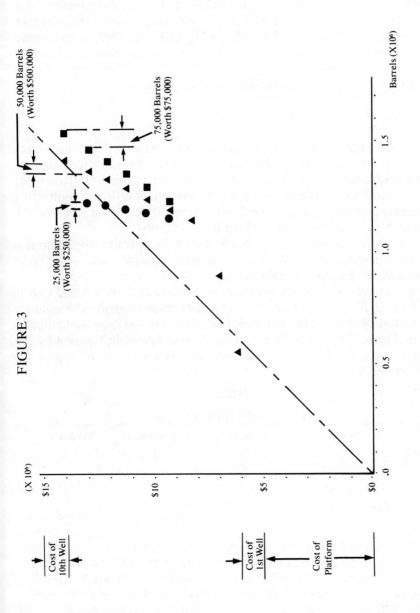

FIGURE 3

On the other hand, a smaller incremental contribution to output and revenue by the fourth and succeeding wells (for example, an increase of only $250,000 in revenue for each $1 million well) would trace a steeper curve such as the one defined by the circles in Figure 3. The addition to revenue would clearly be less than the $3 million residual sacrificed to obtain it, which is another way of saying that the increase in production would be worth less than the bonus bid sacrificed to obtain it.

TRANSFERABILITY OF COMMITMENTS

Clearly the previously discussed simple work commitment bidding system results in more intensive development of the tract to which it is applied. Clear also is the fact that this effect is dependent on the characteristics of the tract to which it is applied, and in particular on the efficiency with which the successive increments of additional expenditures required under the commitment can be put to work to increase output.

One way to increase this efficiency is to allow an operator who assumes a work commitment in the course of acquiring a particular tract to fulfil that commitment through expenditures on a different tract or tracts.

For example, assume that a bidder has acquired both tract A and tract B as shown in Figure 4. If the work commitment assumed in order to acquire a tract must be fulfilled on that same tract, then his maximum commitment on A (Figure 2) is $14 million; and on B (as indicated in Figure 4 by the dashed lines) it is $8 million. Total output from the two tracts will be 2.2 million barrels.

TABLE 1

WORK COMMITMENT BIDDING

	Output (bbl's)	Revenue ($)	Expenditure ($)	Residual ($)
Tract A	1.4	14	14	0
Tract B	0.8	8	8	0
	2.2	22	22	0

Note: All figures in millions

If the operator is allowed to bid on the two tracts jointly or is otherwise permitted to shift a commitment made to acquire one to the other, then his total work commitment will rise to $24 million, with a corresponding increase in output. As shown in Figure 4, this is possible by operating tract A at the point on the output curve which produces the greatest residual and by transfering that residual, as an internal subsidy, to tract B, where, as indicated by the flatter slope of the output curve, it can be utilized more efficiently. The numbers are summarized in Table 2.

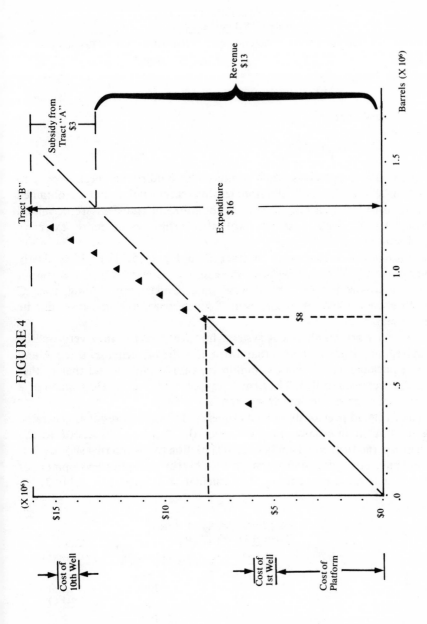

FIGURE 4

TABLE 2

WORK COMMITMENT BIDDING
(Internal Subsidy Allowed)

	Output (bbl's)	Revenue ($)	Expenditure ($)	Residual ($)
Tract A	1.1	11	8	3
Tract B	1.3	13	16	(3)
	2.4	24	24	0

Note: All figures in millions

EXTENT OF DEVELOPMENT

In the above examples work commitment bidding has been shown to result in a more intensive development of tracts than would be obtained under cash bidding arrangements. If commitment transfers are permitted among tracts, such a system will also bring about more extensive development.

Consider the output curve of tract C in Figure 5. Tract C is clearly something of a "dog," because there is no point at which the output function crosses the 45° "breakeven" line. Under cash bidding, tract C would elicit no interest at all; even if given away free, it would not be developed.

Under a work commitment system, however, tract C may very well be acquired and drilled. Assume that a firm has already acquired tracts A and B as a package with a work commitment of $24 million and that neither tract has yet been drilled. The firm is now offered tract C. How much of a work commitment can the firm offer for it?

The firm had previously planned to use the $3 million residual generated on tract A to internally subsidize tract B. If tract C is added to the inventory, the firm could apply to C $1.5 million of the internal subsidy that would have otherwise gone to tract B and thereby make the development of tract C a feasible proposition. The calculation is shown in Table 3.

TABLE 3

WORK COMMITMENT BIDDING
(Internal Subsidy Allowed)

	Output (bbl's)	Revenue ($)	Expenditure ($)	Residual ($)
Tract A	1.10	11.0	8.0	3.0
Tract B	1.05	10.5	12.0	(1.5)
Tract C	0.75	7.5	9.0	(1.5)
	2.90	29.0	29.0	0

Note: All figures in millions

71

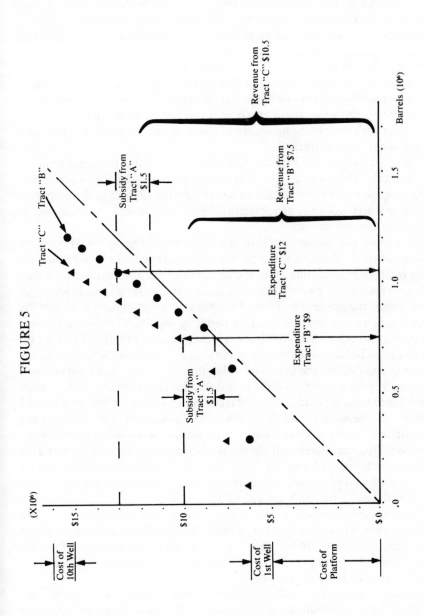

FIGURE 5

Development of the three-tract package under the work commitment system will be feasible with a total expenditure of $29 million. Since he has already been committed to spending $24 million of this, in the course of acquiring tracts A and B, the maximum commitment bid this operator can afford to make on tract C is $5 million.

Besides illustrating the mechanism through which commitment bidding induces more extensive resource development, the tract C example also indicates how the system (with internal subsidies allowed) may work to the advantage of the firms that can acquire the most tracts. Theoretically, a newcomer with no existing inventory of tracts would be unable to make any commitment bid on tract C.

A way of evading this problem would be to make the commitments transferable. This would allow operator Jones to legally assume the obligations to which operator Smith has committed himself in the course of acquiring tracts from the government. Presumably, operator Smith would pay Jones for the favour.

The prospective bidders in these examples have been endowed with the ability to foresee accurately and precisely the output curve associated with every tract. In practice this is not the case. If there is any characteristic that sets the exploration phase of the petroleum and mineral industries apart from other businesses, it is the everyday uncertainty with which its participants must learn to cope. It is perfectly possible, for example, that Smith might acquire tract C with a work commitment bid of $14 million on the mistaken belief that its output curve is that of tract A. After building a platform and drilling the first four wells, the true shape of the curve—and the firm's predicament—would reveal itself. The required $14 million expenditure applied to tract C would leave Smith with a net loss of about $4 million. If he could somehow shed $5 million of the $14 million commitment, Smith would be able to operate the tract with only four wells already drilled and thereby cut his losses to the more acceptable level of $1.5 million. Smith would be willing to pay up to $2.5 million in cash to unload the $6 million obligation, since that is the amount of his maximum additional loss if he can't get rid of it.

Any other operator who is facing or expects to face an output curve with a flatter slope than that faced by Smith will be able to make a mutually beneficial deal with him, since, for the other party, an additional expenditure of $5 million will bring in more than the maximum $2.5 million that Smith will be willing to pay.

Besides putting the small firm in a better position to compete and mitigating the problems of uncertainty for all firms, large and small, a system which allowed the free exchange of work commitments would have the further and more important advantage of maximizing the overall efficiency with which work commitments are utilized. To the extent to which such a market was effective in bringing potential commitment offerers together with potential commitment takers, it would ensure that everyone would be operating at a point where a small increase in expenditure by one operator would produce no more and no less additional revenue than would the same increase applied to any other operator.

A minor but interesting benefit of a market in commitments would be the information it would provide concerning the efficiency with which the work commitment system is eliciting additional output.

This can be understood by applying the concept of the output curve (hitherto used in relation to individual tracts) to the entire universe of tracts being offered under the commitment bidding system. By combining all such tracts and treating them as a single entity, an overall output curve similar to Figure 6 can be drawn. Just as in the case of an individual tract, the point where the curve dips farthest below the 45° breakeven line will be the optimum operating point, the point which results in the resource's largest contribution to economic welfare. The distance between the breakeven line and the output curve at that point is the measure of that contribution, and it is equal to the income that would come to the government landowner were a competitive bidding system utilized. It is this residual—the difference between total revenues and total costs—that the public will be sacrificing to subsidize output.

The efficiency with which the subsidy provided by the work commitment system works to increase output will be a function of the output curve's shape to the right of the point where bonus income would be maximized. Three hypothetical configurations for this part of the curve are shown, and the difference this shape makes to the level of additional output educed by the sacrifice of the residual can be seen. The significance of the price at which commitments change hands will be determined by the slope of the output curve at the point where it intersects the 45° breakeven line. Curves Y and Z both cross at an angle which indicates that, at that point, $1 of expenditure produces $0.50 worth of additional output. In either situation the market price for assumption of a $1 million commitment would be $500,000. Curve X, however, crosses the line at a steeper slope (as shown in the inset), indicating that $1 of expenditure will produce $0.33 worth of additional output. If curve Z accurately represents the overall output curve, the market price for the assumption of a $1 million work commitment will be about $333,000.

From a policy standpoint these numbers, whatever they may be, have considerable significance since they can be used to compare the efficiency of the work commitment system in eliciting additional output with whatever other alternative policies may be available. For example, if the market for

74

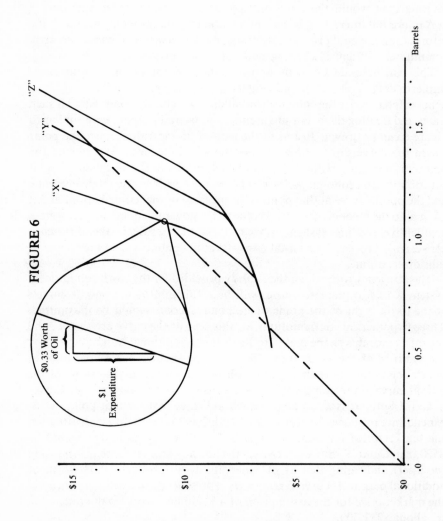

FIGURE 6

commitments indicates that the last dollar of subsidy transferred is producing only 33 cents worth of additional oil, it may very well be that another source of oil development—for example, tar sand or oil shale— could be found that would give a better return.[6]

THE TIME FACTOR

To simplify the discussion of the work commitment system, the time dimension of both expenditures and income has thus far been ignored. Since a dollar in hand today is worth more than the promise of a dollar tomorrow, it has been assumed that all bidders have used a discounting mechanism to take account of the time value of money. Thus, they would reduce all amounts of both expenditure and income streams to their *present value*, that is, the lump sum that the bidder would be willing to receive or give in exchange for the specified income or expenditure stream.

If a work commitment system is to be implemented, it will be necessary to take careful account of the fact that a commitment fulfilled in the next year will have a very different impact on output from one of the same amount fulfilled ten years hence.

If the public is going to give up its bonus income to subsidize output, the time when that subsidy takes effect will presumably make some difference Of course, some arbitrary time limit could be established for the fulfilment of commitments exactly as the United States government sets the five-year term on the OCS oil and gas leases it sells. If free exchanges of commitments are allowed, anyone who wished to distribute expenditures over a time frame incompatible with his commitments could simply enter the market and adjust his inventory of commitments accordingly.

Another way of handling the problem would be simply to apply some appropriate interest rate to every commitment assumed and specify that the amount of expenditure required under that commitment must increase by the amount of the compound interest accumulated in the period between the assumption of the commitment and its fulfilment.

CONCLUSION

The adoption of a work commitment bidding system implies a judgment that existing institutions for private exploitation of public resources result in a suboptimal rate of resource development.

Any argument for the adoption of such a system must first establish that this is in fact the case. Secondly, it must prove that the work commitment approach is the least costly method of achieving the desired higher rate of exploitation.[7]

In comparing various alternative ways of achieving higher output against the work commitment system, some important features of that system are certain to stand out. First, and probably most significant, is the simple fact that the work commitment system results in a subsidy. As such, the criteria for its evaluation should be no less stringent than those applied to a direct appropriation of public funds or a tax concession adopted for the same purpose.

Secondly, the cost of the subsidy conferred under the work commitment approach is impossible to determine *a priori* and difficult of determination after the fact. If there exists a "right" level of subsidy it will be mostly a matter of luck if the foregone public revenue happens to equal that amount. For similar reasons, the benefits of the subsidy in terms of the total increase in output, development, or whatever, are not amenable to accurate quantification. As a consequence, it is doubly difficult to evaluate the system's relative efficiency.

Finally, a properly designed work commitment system allocates the uncertain amount of the subsidy in a fashion that tends to squeeze the maximum additional output from every dollar of subsidy. It does this more or less automatically. This characteristic means that the transfer of resources occasioned by the subsidy needs no affirmative action on the part of policy makers—as does a direct appropriation of public funds—in order to be continued. This fact, however, combined with the intrinsic uncertainty concerning the subsidy's magnitude makes it easier for vested interests to perpetuate such a subsidy long after any real justification for it has passed.

Notes

1. *U.S. Statutes at Large*, vol. 67, p.345. Public Law 212 (August 7, 1953).

2. For an exposition of the conventional industry wisdom on this point, see "Terms for North Sea Oil," in *Petroleum Press Service* 40 (1973): 122-24.

3. In recent testimony before the Senate Interior Committee, industry representatives were unanimous in their opposition to royalty rate bidding. *Outer Continental Shelf Oil and Gas Developments, Hearings Before the Subcommittee on Minerals, Materials, and Fuels of the Committee on Interior and Insular Affairs* (United States Senate, 93rd Congress, Second Session: May 6,7,8,10,1974).

4. Arguments for the work commitment system are expounded in detail in I. White, D. Kash et al., *North Sea Oil and Gas: Implications for Future United States Development* (Norman: University of Oklahoma Press, 1973).

5. Kenneth Dam has touched on some of these matters, however, in "Oil and Gas Licensing in the North Sea",*Journal of Law and Economics* 8(October 1965) and "Pricing of North Sea Gas in Britain," *Journal of Law and Economics* 13 (April 1970). Dam recognizes the subsidy for what it is and concludes that it is unjustified.

6. If it is determined that only a portion of the residual should be applied as a subsidy, this could be accomplished by offering only a portion of the tracts on the commitments bid basis, but allowing the fulfilment of a commitment on any tract acquired from the government. Since the reduction in the subsidy will move the operating point to the left on the output curve (Figure 5) to a point at which its slope is flatter, the result will be a higher assumption price for commitments. The commonsense explanation is that the optimum amount of total expenditure will be reduced only slightly; but that acreage over which commitments that can be fulfilled by that expenditure will be reduced relatively more. Another alternative would be to combine the work commitment system with a royalty, net profits share, or other form of deferred rent collection. This is in fact the arrangement pertaining in the North Sea sector. If the government takes a very high net profit share, the amount of the subsidy will be substantially reduced.

7. There are certain circumstances where a work commitment system could result in no more extensive or intensive development. This would be the case if the commitment were devoted to an activity not contributing to output. For example, if a government awarded a mineral concession to the operator who promised to build the largest smelter, no increase in mine output would result (unless the location of the smelter made lower grade ores profitable to mine).

A New Tax for Natural Resource Projects*

ROSS GARNAUT
ANTHONY CLUNIES ROSS

Mines, as well as land, generally pay a rent to their owner. . . . The metal produced from the poorest mine that is worked must at least have an exchangeable value, not only sufficient to procure all the clothes, food, and other necessaries consumed by those employed in working it, and bringing the produce to market, but also to afford the common and ordinary profits to him who advances the stock necessary to carry on the undertaking. . . . This mine is supposed to yield the usual profits of stock. All that the other mines produce more than this will necessarily be paid to the owners for rent.
David Ricardo

This article proposes a particular form of tax that relates to the rate of return. The article is a discussion of taxing natural resource projects, with special reference to mining, because the tax appears to answer particularly well the special taxing problems of large natural resource projects. There would be practical difficulties in making the proposed system universal for company taxation. Nevertheless, there is no reason in principle why the same method should not be applied to particular projects of other kinds.

The argument of the article depends on certain assumptions about the behaviour of those who make investment decisions for the kind of company that is commonly interested in exploiting natural resources. One is that their investment decisions take account of the projected after tax rate of return on total cash flow and that they set a minimum for the expected value of this rate of return which they will require before investing. The rate of return required may be estimated over the life of the project or over a shorter period (the payback period), or, indeed, there may be different minimum expected rates of return required for the life of the project and for the payback period. These details are not crucial to the argument. In fact, investors may consider rates of return on shareholders' cash flows rather than on total cash flows, but this again is not crucial.

A second assumption is that those who make the investment decisions are subject to risk aversion. This means that the greater the degree of dispersion

*Certain passages and tables in this article are reproduced from the same authors' article "Uncertainty, Risk Aversion and the Taxing of Natural Resource Projects" in the *Economic Journal* 85, no. 338 (June 1975): 272-87; they are published here by kind permission of the editors of the *Economic Journal*.

of the probability distribution of rates of return below the expected value, the higher will be the minimum required value of the expected rate of return, both because the investor will set maximal limit requirements on the probabilities that returns will fall below certain other critical rates (below the required expected rate), and also because there will be a trade off between risk of loss and expected rate of return. At one point a further slightly different assumption is made about the character of risk aversion: that the investor's objective function has a consistently falling marginal rate with respect to returns. It is taken that the government is not subject to risk aversion over its revenue from resource projects and is, instead, simply concerned with maximizing the expected value of revenue.

The minimum expected value of the after tax rate of return for a particular project will be referred to as the *supply price of investment* for that project. This supply price can be expected to vary with general lending and borrowing rates, with the degree of risk expected, and with the investor's pattern of risk aversion. (We believe that the supply price of investment for mining projects of "normal" degree of risk in developing countries of "normal" political stability or instability has been of the order of 15 to 18 per cent in the late 60's and early 70's.) Any returns received before tax in excess of the supply price of investment will be described as the economic surplus or *rent* on the project. This use of the term "rent" is identical with that of Ricardo (as in the quotation at the head of this article), except that we interpret what he calls "the usual profits of stock" to vary with the risk of the investment.

A further assumption in the bulk of the article is that there is no effective competition among investors for rights to exploit particular resources. This assumption is later removed, and consequential modifications are suggested in the tax system proposed. These modifications partially represent a response to some of the other articles in this volume, especially that of Walter Mead on competitive cash bonus bidding.

EXTRACTING THE RENT FROM NATURAL RESOURCES

In the case of natural resource projects, maximizing government revenue from such investments as are made is a particularly critical requirement. It is understandably felt that the benefits derived from possession of the resource, the *rent* in the sense used here, should accrue to the public. Furthermore, industries based on cheap sources of energy or producing minerals or fuels are typically capital intensive, and in a developing country often have limited scope for local purchase of supplies. They may also have undesirable external effects. They may be held to damage the environment and, without compensating expenditure by the government, may contribute to inequality of income between individuals and of facilities between regions. Thus, particularly in developing countries, any benefit from them must come mainly through public access to their profits, and the central task of economic management in these industries is to maximize their contribution to government revenue. This task has two aspects: ensuring that the resources are used in such a way as to increase social product to the greatest extent possible, and capturing for the public as large a proportion as possible of the benefits so generated.

The former of these requirements may in many situations entail management of the operation by a private firm which has at least a substantial share of the equity in the project. We shall argue that, provided this method of exploitation is adopted, the tax proposed here appears to be capable of securing for the government a larger expected absolute amount of rent from each resource than other taxation systems applied to projects *ex ante*. We shall also argue that it does this with as little tendency as (or less tendency than) any other tax systems to reduce efficiency in the use of resources, at least in a national economy that relies on external finance for its major resource projects. (In an economy in which major resource projects are not mainly financed externally, there is arguably an element of incentive to allocative inefficiency introduced by the tax in its pure form; but this can be overcome, though not necessarily without costs in other directions, by applying the tax in a hybrid form with normal company profit tax.) The only tax systems that could be expected to secure for the government a larger share of rent from particular projects would be those that depended on *ex post* adjustments to the revenue terms on which the investments had been made. Such adjustments, by increasing the uncertainty and therefore the risk associated with future investment, would reduce the revenue to be expected from new projects and from extensions of existing projects.

Throughout the article the term "expected" is used in a statistical sense. The *expected value of revenue* is the mean of a probability distribution of possible revenue outcomes. A tax system which, by comparison with another, promises a higher expected revenue in this sense may, of course, actually turn out to have given less revenue in the case of a particular project or indeed over a range of projects. It is assumed here that significant numbers of potential resource projects are available, so that it is reasonable to consider revenue expectations that various systems might give over a range of projects.

It is a natural aspiration for governments to appropriate the whole rent of resource projects. There are various reasons, however, why this cannot be satisfactorily done under private management. To try to do so with a 100 per cent marginal rate of tax on profits would remove any incentive to efficient operation. To apply licence fees or royalty charges *ex post* which were designed to have the same effect as a 100 per cent tax on profits would presumably remove incentives to efficiency in precisely the same way once the general character of the policy was realized. Licence fees or royalties could be applied *ex ante* to capture 100 per cent of the rent if there were certainty about output, costs, and prices; but under real-world conditions of uncertainty about all these variables this, too, is not a practical possibility. Collecting the whole of the rent *as a general rule* thus appears to be practically out of the question.

Can one then have systems whose expected revenue yield equals the expected value of the rent? This is possible in principle. If, however, the tax system imposed increases the risk of loss—if it increases the spectrum of probabilities that the rate of return (over the life of the project or during the payback period) will fall below certain critical levels—the investor must react by raising his supply price of investment and so reducing the rent available to be tapped. Each of the common types of tax (fixed licence fee, specific royalty, *ad valorem* royalty, proportional profit tax, and the kind of progressive profit tax that relates the rate of tax in a year to the ratio which that year's profit bears to some estimate of funds, or shareholders' funds, invested) increases the risk of loss in the terms described. (Even a proportional profits tax, or a progressive profits tax, insofar as it may be collected before loans have been repaid and shareholders' investment recovered at an appropriate rate, increases the probability that acceptable rates of return will not be realized.) Hence, even if such taxes could be devised to collect an expected amount equal to the expected value of the rent, their effect on the risk in the investment would reduce the rent available to be collected.

We can express this problem another way. In negotiations over the tax system applicable to a project, the investor will inevitably act as if he took a pessimistic view of the investment's prospects and he will thus require an expected after tax rate of return considerably higher than he would be prepared to accept if there were no risk of loss, and one that is higher than he would accept if the tax system itself were not such as to increase this risk. Generally, the kinds of tax that relate actual revenue yield least closely to the before-tax profitability of projects will entail the greatest risk of loss for any given expected tax yield. In this respect a fixed annual or once-for-all licence fee is the least satisfactory and profits tax the most satisfactory of the kinds listed in the previous paragraph. The rent available to be collected by means of the former will therefore be lower than that available by means of the latter.

The problem of taxing adequately through systems of the kind listed (either singly or in combination) is then, in the first place, one of uncertainty, of imperfect knowledge on the part of both government and investor, and it would remain even if their knowledge and capacity for prediction were equally good. If there is any serious risk that the project before tax will be close to the margin of acceptable profit, the investor will presumably require licence fees, royalties, or proportional profit taxes to be applied at fairly negligible rates. If, as may well happen, results are far better than the investor fears, the actual taxes collected will be low in relation to the profits earned.

The problem is increased if the government's knowledge of the probabilities of various levels of cost, price, and output, and of the supply price of investment is inferior to the investor's. In that case, the government may actually underestimate the revenue that the investor is in the last resort prepared to offer. The bargaining situation, which in the absence of competition is inevitable if reasonable attempts are to be made to maximize the revenue through any of the kinds of tax mentioned, is one in which the quality of the government's market information and negotiating skills are critical to its success. It is not impossible that on both scores the representatives of a large multinational investor will be superior to those of a small developing country.

There is a superficial attraction in applying more conventional forms of tax *ex post* to each resource project at rates which secure a high proportion of the rent for the government. However, such systems add taxation uncertainties to the natural uncertainties of projects and so raise the supply price of investment, unless the principles of *ex post* adjustment are declared, in which case the problem is identical with that in *ex ante* systems.

The Resource Rent Tax (RRT) described in the next section represents an attempt to overcome the two problems mentioned: not to increase the risk of loss above that which would obtain in the absence of taxation and hence to maximize the rent available to be taxed; and not to leave the revenue arrangements to negotiations in which a government may well be at a disadvantage through relative ignorance.

THE RESOURCE RENT TAX

We shall describe this tax first in a pure or extreme form and later introduce modifications which may be useful in practice.

The RRT is briefly a profits tax that begins to be levied, and at a very high rate, when a certain threshold rate of return on funds invested (as measured by discounted cash flow methods) has been realized. The rate of return is calculated on the total cash flow of the project, not simply on the shareholders' cash flow.

Assessment for RRT is based on the stream of annual *net assessable receipts* from the beginning of expenditure on the project. Net assessable receipts for each year represent the excess of assessable receipts over deductible payments.

Assessable receipts include all receipts of the operating company other than receipts in the nature of provision or repayment of capital. They thus exclude the value of loans and shareholders' funds received and the value of loans repaid to the company by any other entity. They include receipts from the sale of depreciated, obsolescent, or other assets formerly purchased for use by the company. Interest and dividend receipts would be excluded, but, for reasons to be explained below, the rules must prevent a business taxed by RRT from *receiving* either dividends or interest.

Deductible payments are defined similarly as covering all payments by the company other than payments in the nature of repayment of capital, provision of capital, and rewards for the provision of capital. They thus exclude repayments of loans, payments of interest, dividends and bonus issues, and provision of capital. They do not include past payments of RRT. They do, however, include payments of any tax other than RRT.

With a few exceptions that can be specified, the definitions of receipts and payments used in the assessment of company income tax can also be used in the assessment of the RRT. The most important exceptions relate to the special treatment of interest payments and depreciation described above. Increases in stocks of unsold raw materials will not be assessable receipts, and it will be necessary to specify the time at which sales become assessable receipts. Either the time of physical export or the time of physical receipt by the purchaser would be appropriate. The income tax definitions of deductible payments will need to be limited, in the case of payments in respect of contracts, to payments for services rendered in the relevant accounting period. Deferment of liability for taxation through early payment under longer-term contracts for services occurs with conventional income tax as well, especially in transactions among affiliates, but the cost to the revenue of early payment will be greater with RRT.

Prospecting and investigation expenses may be counted as costs in industries such as mining, where they can readily be attached to particular projects. Where this is not readily done, as in petroleum extraction, these costs should be ignored and a correspondingly higher threshold rate applied.

Assessment proceeds by calculating the annual accumulated value of net assessable receipts, at a discount rate, x per cent, corresponding to the supposed supply price of investment in all the circumstances of the project. When a year is reached where the accumulated value is positive, that value forms the tax base for that year and is taxed at a very high rate of a per cent. Thereafter, so long as net assessable receipts are positive, such receipts in each year are taxed at a per cent. If in any subsequent year the net inward cash flow is negative, no tax is collected in that year and the calculation of accumulated value resumes from that year until once more a positive value is recorded. Once more the positive accumulated value serves as the tax base for the year in which it occurs, and net assessable receipts are taxed at a per cent in subsequent years for as long as they remain positive.

Expressed in this form, the RRT can be regarded as a company income tax with:

a. no deductions for interest payments;
b. immediate 100 per cent deduction for capital expenses;
c. unlimited carrying forward of losses discounted for time;
d. rates of tax considerably higher than those of most company tax systems.

If it were desired to tax beyond higher profit rate thresholds at higher rates than a per cent, the process of discounting to find net present value for each year could be repeated with the higher threshold, y per cent, as the rate of discount. No deduction would be made for RRT assessed over the previous threshold. An additional tax of b per cent would be applied to any positive accumulated value discounted at y per cent.

The total effect of the system would be to tax company returns in excess of x per cent (after company tax and royalties, if any) at a rate of a per cent, and returns in excess of y per cent, at a rate of (a + b) per cent. If it were thought desirable, further gradations could be introduced so that returns in excess of z per cent could be taxed at a rate of (a + b + c) per cent. The accumulation of net assessable receipts at a given discount rate ensures that the timing of a company's income and expenditure has no distorting effect on the internal rate of return at which it becomes liable for RRT.

Table 1 shows a hypothetical example in which accumulated value at a 10 per cent discount rate is taxed at 50 per cent and accumulated value at a 20 per cent discount rate is taxed at an additional 25 per cent.

Because it is based on revealed profitability, the RRT to a large extent prevents the company from exploiting the government's relative ignorance. Whereas alternative taxation mechanisms require the government to judge both the probability distribution of costs and prices *and* the supply price of investment, only the latter judgment is required for RRT purposes.

If the problem were to fix the rates for one project only, values for x and y and for a and b would be chosen to maximize the level of tax thought to be consistent with the particular company's decision to invest. If a general system were desired, either for all resource projects or for each industry, the rates would probably be set at levels expected to maximize overall tax receipts resulting from projects. This would, of course, entail a compromise between too lenient rates and those severe enough to discourage an undue proportion of potential projects. The rule might be modified if it were thought that royalties and other similar charges made inadequate allowance

TABLE 1
HYPOTHETICAL EXAMPLE OF RESOURCE RENT TAX

(1)	(2)	(3)	(4)	(5)	(6)	(7)	(8)	(9)
Year	Assessable receipts	Deductible payments	NAR = (2)-(3)	Accumulated value of NAR of current year and previous series of years with negative current or accumulated value of NAR (10% discount)	Tax on returns over 10% threshold at 50% rate of tax	Accumulated value of NAR of current year and previous series of years with negative current or accumulated value of NAR (20% discount)	Tax on returns over 20% threshold at 25% rate of tax	Total Tax = (6)+(8)
1	–	100	-100	-100	–	-100	–	–
2	–	300	-300	-410	–	-420	–	–
3	50	100	-50	-501	–	-554	–	–
4	200	50	150	-401	–	-515	–	–
5	200	50	150	-291	–	-468	–	–
6	200	50	150	-170	–	-412	–	–
7	200	50	150	-37	–	-344	–	–
8	200	50	150	109	54.5	-263	–	54.5
9	200	50	150		75	-166	–	75
10	200	50	150		75	-49	–	75
11	200	250	-50	-50	–	-109	–	–
12	200	50	150	95	47.5	19	4.75	52.25
13	200	50	150		75		37.5	112.5
14	200	50	150		75		37.5	112.5
15	200	50	150		75		37.5	112.5

for the general external diseconomies of resource investments. In that case, rates and thresholds might be set in such a way that they deliberately discouraged projects which, in the absence of tax, would be close to the margin of expected profitability.

Rates of taxation could be extremely high in the later years of a highly profitable project, without either encouraging the running down and disposal of assets or discouraging additional investment in exploitation of the natural resource. (In this, the RRT would differ from company profits tax as usually assessed). This is because any sale of assets would be counted as a receipt for RRT purposes, so that its value to the company would be reduced by RRT in the same proportion as profits. Similarly, the cost of new investment or maintenance expenditure would be reduced by the RRT in the same proportion as income earned on the investment is reduced. However, although a very high rate of tax on marginal profits would not promote disinvestment through the effect on returns to investment, the marginal rate would, of course, need to be kept below 100 per cent to maintain the investor's interest in the efficient management of the project. This RRT characteristic of not reducing incentives to investment is the basis for the earlier assertion of its superiority on grounds of efficiency. A profit tax with immediate 100 per cent depreciation and unlimited carrying forward of losses would, of course, imitate this advantage and would have many of the features of a RRT with zero per cent threshold.

The RRT as defined above is based on rates of return on total cash flows for the project. It could be adjusted to vary with the rate of return on shareholders' cash flows. However, we favour using total flows, since this method avoids arbitrary and unintended effects on the choice between equity and loan finance.

PRACTICAL CONSIDERATIONS OF ADMINISTRATION

As with any income tax, there has to be careful surveillance of purchase and sale prices to prevent prices used in intrafirm transactions from being used as a means of tax evasion, or of shifting tax liability to another country or to a firm taxed in a more lenient fashion.

The main special requirement of RRT involves defining the entity that is taxable under this system. Since the RRT (even as defined in pure form above) is a progressive tax (in having at least two rates: zero and, say, 80 per cent), there would be opportunities for tax evasion if the investor could spread profits for tax purposes to and from other entities. Thus, if RRT is to be applied to a mining project, the entity to which it is applied must be limited to certain definite processes carried out in a particular place: simply mining, or mining and certain types of processing, for example. Other activities would have to be undertaken by affiliates separately taxed (in most cases presumably by the normal system of income tax applying in the country). This would apply to any earning of interest on idle funds. These funds would have to be lent free of interest to an affiliate taxed separately. Similarly, there would be a prohibition on holding of shares in other companies by the entity taxed by RRT.

The only likely administrative disadvantages of the RRT as compared with ordinary income tax are that it is a new and additional system and that it introduces the special problem of defining the taxable entity. Records to be kept are little different and in some respects less complex than under ordinary income tax. Removal of considerations of depreciation is one important simplification.

MODIFICATIONS

If the "pure" form of RRT with one threshold and one tax rate were modified by introducing further progression, this would be one way of meeting the problem of governmental ignorance of and variation in the supply price of investment. If the government is confident that the supply price lies somewhere between say 10 per cent and 15 per cent, then it can begin collecting at a "normal" company tax rate (say 50 per cent) when the project has attained a 10 per cent threshold rate of return, and move to the higher tax rate (say 80 per cent) when a 20 per cent threshold before tax (that is 15 per cent after tax) rate of return has been reached.

If, because of risk aversion, the investor's utility function rises consistently at a diminishing marginal rate with returns, then progression of rates of tax through various rates of return also tends to increase the expected revenue consistent with a given expected utility for the investor.

One problem with RRT is that the flow of revenue is likely to be less regular and predictable than in the case of ordinary proportional profit taxes or (a fortiori, in ascending order) ad valorem royalties, specific royalties, or fixed annual licence fees.

If there are political or budgetary difficulties about the possibly long delay before revenue is realized, then the first threshold rate may be lowered well below what anyone supposes to be the supply price of investment (even to a negative rate). If the rate of tax that comes into effect above this threshold is moderate, some of the essential advantage of RRT may be retained.

Alternatively, to meet the same difficulty, RRT could be combined with a moderate rate of company tax or royalties or both, these other taxes being treated as costs for RRT purposes. If, moreover, there is any doubt whether RRT will be treated as income tax for purposes of international double taxation or tax sharing arrangements, this device may avoid the loss of benefits under such arrangements. A compromise would be to tax by ordinary company tax or RRT in any year, whichever is the higher, and to offset the company tax payments of early years, insofar as they exceed RRT liabilities, against future RRT liabilities, insofar as they exceed company tax liabilities. Companies might be given the option of choosing this hybrid system, rather than pure RRT, in case they may derive any overseas tax advantages from doing so that will outweigh for them the drawbacks of the hybrid system. The hybrid system would reduce the tendency to fluctuation of revenue receipts that would exist under pure RRT, and it would avoid what might sometimes be embarrassingly long tax holidays under RRT.

Thresholds applied *ex ante* to each project might be related to some standard international borrowing rate so as to reflect the interest element in the supply price of investment and, in so doing, make some allowance for expectations of inflation.

It would be useful, in the absence of competition for particular exploiting rights, to have, at any one time, a single set of RRT rates and thresholds for new investments in each major resource industry; for example, one for mining, one for petroleum and natural gas, one for forestry, one for smelting and refining based on cheap hydro power. This would remove the necessity for tax bargaining with individual investors (in which a government may not be particularly adept), and it would also have the great advantage that investors would know, from the beginning of prospecting or other investigation, just what the fiscal conditions of their operation would be if they decided to go ahead. Such a uniform system would not be appropriate with other kinds of tax, the rates and timing of which must be tailored to individual projects if there are generally to be reasonable revenue prospects.

The supply price of investment is raised if the investor doubts the stability of the rate at which the host country's currency exchanges for his own or for others he is interested in. This problem is avoided if all calculations are made in the currency of the investor or in a currency nominated by the investor. The only requirement is that the same currency be used throughout the calculations based on conversions at the ruling exchange rates at the time transactions are made.

It may be considered politically preferable to provide that the investor should be liable (if required) to transfer equity free of charge to the government on attainment of a certain threshold rate of return, rather than move to a higher rate of tax. Fairly close equivalences between these two kinds of liability can be worked out. If the government has the right to take a majority holding, when, but not until, a 20 per cent before tax (or say 15 per cent after tax) rate of return has been realized, it may not even be necessary from the investor's point of view to construct the kind of management contract which is usually made when the government acquires a majority holding from the beginning of a project's life.

MODIFICATIONS FOR THE CASE OF COMPETITION

The discussion hitherto has assumed that there is no effective competition for mining, drilling, or logging rights. Where effectively competitive bidding is possible, it seems desirable to exploit its advantages as explored in Walter Mead's article in this volume. His evidence from bidding for United States offshore oil suggests that, where there is significant competition for petroleum drilling rights, the highest bidder tends to be excessively optimistic, with the result that cash bonus bidding appears to leave the successful bidder with a lower rate of return, after the cash bonus has been paid, than one could reasonably *plan* to allow him under RRT. Tussing's article in this volume makes the point that the highest bidder is not the average or representative bidder. He tends naturally to be the one most optimistic about the prospects of the particular resource for which he is bidding. Thus, where competition is at least a possibility, there appears to be a strong case for some form of competitive bidding. On the other hand, where competition cannot be entirely relied upon, or where exceptionally high after tax returns, even on one individual project, are likely to have adverse political effects, there seems to be good ground for applying the realized profitability tax principle embodied in the RRT.

Two methods of combining these approaches to get the benefits of both suggest themselves. One is to have competitive cash bonus bidding and an RRT with fixed rates and thresholds, the cash bonus being treated as a cost for RRT purposes. The other is to apply the RRT but allow competition through bidding for one of the RRT thresholds. In the latter case, the government should set a maximum rate of return for, say, the upper threshold, then, allow investors to bid threshold rates below or equal to that maximum, with the rights going to the investor who bids the lowest threshold.

The former method of combination could well be suitable for petroleum and natural gas allocation in an established field. Here, investigation and extraction depend substantially on the same investment. Whether a government should opt for cash bonus bidding, for an RRT-type tax, or a combination of the two would depend on judgments about the effectiveness of competition, the relative importance of overoptimism and risk aversion, the quality of profit assessment for tax purposes, and the strength of political objections to exceptionally high rates of after tax profit on particular projects. If competition were possible, but the overoptimism effect were thought relatively unimportant, it might be considered preferable to abandon the cash bonus bidding element and to have bidding for RRT thresholds.

The mining of nonferrous metals involves a different problem. Investigation can be separated from extraction, and it would not be desirable to have a high charge on investigation. On the other hand, the prospector must have a priority claim to mine if there is to be an incentive to private exploration. In normal circumstances the prospector may commonly be the only applicant for the mining lease. In that case asking him to bid would be pointless.

The only satisfactory way to introduce competition at this point would be for the government to conduct all the exploration itself and to publish any positive results. The problem would then become similar to the oil lease case, and the same factors would affect the relative merits of cash and bonus bidding alone, a combination of cash bonus bidding with fixed RRT, fixed RRT alone, and RRT with threshold bidding.

It is likely, however, that in this case the pure gambling element would be less important than for oil leases, and any cash bids would be correspondingly less widely spread. As a result, the government would be unlikely to get much benefit from overoptimism, and the RRT with threshold bidding (or a tax embodying a similar principle) might be the best solution.

It might be possible to introduce competitive bidding with nonferrous metal mining by calling for bids for prospecting rights whenever an application is made to prospect in any area. The successful bidder could subsequently mine if he chose to. This system would require that the area be given a prior clearance for mining on environmental and local-political grounds. In these circumstances, cash bonus bidding by itself would fairly clearly be inadequate, but bidding for RRT thresholds would make sense if there were in fact competitors.

We do not suggest that either of these possibilities for applying competitive bidding in the case of mining is necessarily desirable. Where there is competition, however, threshold bidding seems likely to have advantages over cash bonus bidding in mining applications. On the other hand, cash bonus bidding, with or without fixed RRT, may provide the best option when competition is available for petroleum or natural gas in established fields.

BALANCE OF ADVANTAGE

The principal benefit of the RRT arises because those taking decisions for an investing company have to guard against the risk of very low returns. The tax arrangements an investor will accept will therefore reflect a pessimistic assessment of the outcome of the investment. If a project turns out to be even moderately successful, and attains, say, the *expected* value of the rate of return before tax, the revenue fixed according to alternative taxation systems will seem meagre and the after tax returns to the investor correspondingly great. Apart from all else, this is a potential source of political problems to the investor and to the government that has permitted it to operate under these conditions. The RRT avoids this problem by relating returns strictly and entirely to realized profitability.

On this score, at least in the absence of potential competition for rights, the RRT appears superior to other forms of tax applied *ex ante*, while prescribing taxes *ex post* without a prefixed rule suffers from the enormous drawback that it introduces a further large element of uncertainty into the investor's calculations.

The advantages of RRT (with fixed rates and thresholds for each industry) over other systems as a means of taxing privately owned natural resource projects in the absence of effective competition for rights are that it can be constructed to give a higher expected revenue yield for each industry, particularly when there is a differential ignorance on the part of the government; that it minimizes the tax disincentive to new investment in existing projects; that it saves administrative resources that would otherwise be devoted to fiscal negotiations; and that it virtually avoids the political embarrassment that may arise if excessively high after tax returns accrue to private resource investors.

Its disadvantages over the other systems are the greater variance in the probability distribution of total revenue flows from particular projects and industries, certain administrative costs associated with running a new system of income tax and of fixing boundaries between the scope of general and special systems, and the political objections to what may sometimes seem to be excessively long tax holidays.

This suggests that the RRT's balance of advantage will be greatest when individual resource projects are relatively important in revenue generation, when there are resource projects in a variety of different industries, when governmental access to business and marketing intelligence and governmental capacity in negotiating are relatively weak, and when direct tax administration is relatively strong, or is capable of being made so.

Comment

GLENN P. JENKINS

In their pursuit of revenue from mineral enterprises, governments have resorted to a wide variety of taxes and levies. They can be organized into four broad categories: fixed charges such as rents, fees, and licence charges; royalties and other payments levied on the amount or value of the mineral extracted; taxes levied on the income earned by the enterprise, personal or corporate income taxes, withholding tax, or any other type of income tax such as an excess profits tax; direct equity participation by the government in the mineral corporation.

Before discussing the specific proposal of the Resource Rent Tax (RRT) developed previously, I wish to point out the advantages and disadvantages inherent in the four broad procedures of revenue collection outlined above. In this way, we will have a better understanding of the attractions and pitfalls inherent in the specific scheme for mineral taxation that has been outlined by Garnaut and Ross.

RENTS, FEES, AND LICENCE CHARGES

These charges are usually levied as part of the payment to obtain the right to the use of mineral lands, to carry out exploration, or to operate an enterprise within a jurisdiction. The obvious advantage to these forms of raising government revenue is the administrative ease of their collection. As they are usually a fixed charge per period, then, if they are not so large as to prevent the opening of a mineral body for production, they will not alter the marginal variable costs of production and thus not distort the production decision of the mineral enterprise. When combined with a royalty on production to generate a target amount of government revenue, the existence of the fixed fees or rentals per period will enable a lower royalty rate to be set, hence lowering the distorting influence of the royalty on operating decisions.

The disadvantage of such fixed charges is that they are seldom related to the quality of the mineral deposit; therefore, they may distort the decision to begin or not to begin production. Also, since they do not vary with the annual profitability of the project, they will increase the risk of the project, where risk is measured by the variance of the annual returns. Because of these features, fixed charges generally will not be effective instruments for the collection of a large proportion of the economic rents that may exist in the mineral sector. Nevertheless, rental and licence fees have constituted approximately 25 per cent of the total government revenues from oil production in the province of Alberta, Canada. In this case, these fixed charges have been combined with a scheme of bonus bidding for the sale of mineral rights and a royalty that has been approximately 16 per cent of production. [1]

ROYALTIES

Royalties are usually levied as a fixed percentage of the gross value or physical amount of mineral production. In certain situations, such as in the Middle East oil producing countries, while *de facto* royalties are collected, they are called income taxes because they are applied to a previously fixed amount of profits per barrel. In this discussion, any tax levied or negotiated as a fixed amount per unit of output prior to production is treated as a royalty. The principal advantage of a royalty as a form of taxation is the ease with which it can be collected. When a government is attempting to levy an income tax on the profits of a corporation, it may have great difficulty determining the true amount of profits from the investment. The widespread use of transfer prices between affiliates of integrated multinational corporations, along with the absence of reliable auditing procedures of financial statements in many countries, prohibits the use of genuine corporation income or excess profits taxes. On the other hand, the amount of output from the operation can ordinarily be easily identified; thus the government can readily measure the base on which the royalty is to be levied.

The principal drawback of a flat rate royalty on the production of a mineral is that it is levied on costs as well as economic profits. Such a tax will often lead to bypassing high cost ores in the production from a mineral body that would have been utilized in the absence of the tax. Thus an economic loss is inflicted.

The high-grading effect of a fixed rate royalty can be somewhat overcome by using a variable royalty rate which will move inversely with the costs of production from a mineral body. This is especially important in the case of a mining operation where the quality of the ore may vary greatly within a given mine. While a substantial royalty may be appropriate for the high grade ores, it would cause production from some lower grade deposits to be unprofitable, consequently these lower grade ores would be wasted. In this case, multiple royalties would be required in order to have economic efficiency even for the production from a single mineral body.

A system of variable royalties requires much of the same information as that needed for an effective business income tax. As in the case of the business income tax, estimates are needed for the costs and revenues for each property before a calculation can be made of the available economic rent to be captured by the royalty.

INCOME TAXATION

When the accounts of a mineral enterprise accurately reflect the private costs and revenues of its operation, then a tax levied on the firm's net income has a number of advantages over other forms of taxation. A tax based on income or profits will avoid the problem of having to estimate the costs of extraction and the price of the output before the period of production, as would be necessary if a variable royalty were levied. The income tax also eliminates the problem of high grading, since there will always be an incentive (when the firm is operating competitively) to use the poorer grades of ore up to the point where the marginal cost of production equals the price of output.

When a mineral body is being developed by a corporation which is a foreign affiliate of a multinational corporation, the tax policies of the country of the parent corporation will be an important factor in determining the attractiveness of the income tax on mineral production in the foreign country. For example, affiliates of United States parent corporations are allowed to use income taxes paid to foreign countries as tax credits against the United States tax liabilities due on this foreign earned income. [2]

The United States foreign tax credit is limited to the amount of the United States tax liability on the foreign income alone. There are two ways to calculate the limitation on the amount of foreign tax credits that can be used to offset the United States tax liability on this foreign earned income. First, there is the per country limitation when a separate computation is made of the United States tax liability and the available tax credits for each country where income is earned. Alternatively, the taxpayer may irreversibly elect to use the overall limitation method under which the United States taxable income (and losses) from all foreign sources are pooled, as are all foreign taxes. One aggregate computation of the United States tax liability is then performed using this pooled data. Under both types of limitation, creditable foreign taxes in excess of the limitation may be carried back for two years or forward up to five years and added to creditable taxes in a year in which there is a shortage. [3]

By being able to credit foreign taxes against the home country's income taxes, a multinational corporation can pay income taxes to foreign countries up to the level of home country income taxes due on this income without increasing the total worldwide tax burden of the multinational corporation.

The home countries of most multinational mineral corporations have allowed a wide range of taxes to be used as creditable foreign incomes taxes. In the case of United States oil companies operating in the Middle East, the taxes paid to these countries are allowed as foreign tax credits to offset the United States income tax liability on the foreign income, even though the foreign taxes are calculated on artificially constructed "incomes" having little or no relationship to the true profitability of the foreign affiliates.

Because of the overall limitation method of calculating available foreign tax credits, any foreign tax credits from one country in excess of the home country's income tax liability can be used to offset the home income taxes due on foreign earned income from another foreign country. Typically an international mining or petroleum company pays large enough amounts of creditable but unavoidable foreign income taxes (which are often in reality royalties) to have an overall excess of foreign tax credits. In these circumstances, the multinational firm has an incentive to transfer profits from countries where their affiliates are facing a high marginal tax rate to either tax haven countries or to areas with a low marginal tax rate. For the international petroleum industry, it can be shown that in most foreign producing countries, while the average tax rate on income from crude oil production is substantial, the marginal tax rate is zero. This occurs because the income taxes in these countries are calculated as an amount per barrel of crude oil produced, not on the amount of true economic profits generated by the production. Therefore, these petroleum companies have had an incentive—one which they have acted upon—to transfer taxable income from their downstream activities located in the consuming countries to the producing countries. In this way they can decrease income taxes paid to consuming countries such as Canada, Europe, and Japan, where the income tax varies with recorded profits, and yet not increase taxes paid to the producing countries. [4]

This transfer of income between affiliates is facilitated by the integrated nature of mining and petroleum corporations. The vertical integration of activities often extends from primary production through the transportation sector to the final sale of the product. This leads to a situation where there is an absence of market prices for many of the products traded between affiliates; hence, shifting taxable income between countries by transfer pricing is easy to execute.

For a country to stop these transfers of taxable income, it must estimate the correct values of the costs and revenues for each firm. The problems that tax authorities would have in constantly monitoring and correcting the accounting data for a set of mining and petroleum corporations are difficult to exaggerate. Before a government embarks on a policy of taxing the mineral industries by a tightly controlled income tax system, it should compare the economic loss that is created by the administrative costs of enforcing such a scheme with the economic inefficiency of other methods of mineral taxation.

EQUITY PARTICIPATION BY GOVERNMENTS IN MINERAL ENTERPRISES

In recent years, governments of mineral producing areas have tended to insist upon owing a part of the equity of the mining or petroleum enterprises operating within their jurisdiction. In some cases, governments have obtained a share of the equity of firms for amounts less than the normal market prices of the stocks. We can view the profits which accrue to the government's shares obtained in this manner partly as a profits tax on the income generated by the private sector's investments.

Some governments see equity participation as a way, on the one hand, to monitor the activities of the mineral enterprise to ensure that citizens receive training in the operation of the firm and, on the other, to collect revenue from the mineral production. While this approach may be very advantageous for a country planning to take over the operations of the firm eventually, there are some serious drawbacks inherent in minority equity participation agreements.

Unless there is a very close surveillance of the prices used in the accounting of interaffiliate transactions, the minority shareholder (here the government) may find itself receiving less than its share of the total income generated by the enterprise. The Middle East countries with participation agreements for petroleum production reduced the incentive for the international petroleum companies to transfer income out of their ventures by contracting for their portion of the "profits" of the activity in terms of a fraction of the barrels of oil produced, rather than a percentage of accounting profits.

Although equity participation has sometimes been considered a way in which a government can effectively control foreign operated enterprises, it can easily become just another expensive way to collect an income tax.

RESOURCE RENT TAX

The tax system proposed by Garnaut and Ross is essentially an income tax (applied to the mineral extraction sector) with a zero rate up to the point where a threshold rate of return is being earned on the investment and a very high marginal tax rate on income in excess of this threshold rate. As a tax based on accounting profits, it has all the advantages as well as the drawbacks of the income tax discussed previously. However, this tax would provide a much larger incentive for corporations to shift income into tax haven countries or engage in wasteful practices than would an ordinary corporation income tax designed to collect the same amount of revenue.

To illustrate, assume that a mining corporation with 100 per cent equity is earning a gross of tax rate of return on its investment of 20 per cent and is paying a corporation income tax of 40 per cent. In this situation, the corporation earns a net of tax rate of return of 12 per cent and pays corporation income taxes each year equal to 8 per cent of the value of the investment. Now let us assume that the threshold rate of return allowed by the RRT is 10 per cent and the government again wishes to collect an amount of revenue equal to 8 per cent of the value of the investment. To do this, it will have to set a marginal income tax rate equal to 80 per cent on the RRT base in order to collect the same amount of tax revenue as with the ordinary corporation income tax of 40 per cent. In this example, the incentive to avoid taxation by transfering taxable income out of the country has doubled. Alternatively, the management of the corporation may find it worthwhile to decrease profits by superfluous business expenses which they enjoy as a substitute for earning profits, since now the government will bear 80 per cent of these additional expenses through reduced tax revenues.

While this tax system attempts to allow for a normal rate of return on investment and then to tax only the excess profits, there are serious difficulties in determining what is the relevant investment base on which to apply the threshold rate of return. To enable a mineral enterprise to survive through time, it must either engage in exploration and development expenditures or purchase proven ore bodies or reserves. If it does its own exploration and development work, then for every successful mine there will be also many failures. Garnaut and Ross would not allow these expenses to be included in the investment base of an operating mine, but instead, suggest raising the threshold rate of return of all successful mines. Increasing the threshold rate of return for the mineral extraction sector does not change the basic problem: if mining companies engage in unsuccessful exploration activities the government does not share in these expenses, while at the same time it may be taxing the existing operating mines on their "excess" profits.

The reaction of a mining industry to this set of tax incentives will likely be to divide its operations into two parts: an exploration and development sector and a mineral extraction sector. If the exploration and development sector is taxed under a normal corporation income taxation system, it will expense or depreciate its costs and sell its proven properties to the mineral extraction branch. The fair market price of these properties will already have capitalized into them any of the economic rents that exist. Thus, the exploration and development sector will be taxed on its profits including the economic rents by a corporate income or capital gains tax system, while the mineral extraction sector will be taxed by the RRT. The government's problem of taxing economic rents in the mineral extraction industry will now be solved, because no economic rents will exist there to be taxed under the RRT system.

The difficulty of determining what is the relevant investment base on which to calculate the threshold rate of return is aggravated by the existence of working capital. While Garnaut and Ross say that they would not allow "idle funds" as part of the investment base, nevertheless any mine operating as a going concern must have a significant amount of working capital which may, at various times, include securities as well as cash and inventories. Depending on the investment plans and payments structure of the firm, these different kinds of working capital are unlikely to remain in a constant proportion to the fixed capital stock.

A novel feature of the RRT, as designed by Garnaut and Ross, is that it would allow for a complete recapture of the present value of the invested capital of the firm before the government would begin receiving any tax revenues. In most cases, this would entail a wait of five years or longer before any tax revenues are collected. Few governments could politically afford to give what would appear to be a tax holiday for such a period of time.

Any tax system applied to the mineral industries will only be as efficient in its capture of economic rents and in avoiding the creation of economic distortions as the quality of information available to the tax authorities allows it to be. The same type of information is required to successfully implement the RRT as is needed for most income tax or variable royalty systems. While the RRT is an interesting attempt to construct a tax system which has a minimum influence on the risk of a mineral extraction project, the economic distortions that it would create with its very high marginal income tax rates and administrative complexity make it an undesirable way of taxing the mineral industries in most situations.

Notes

1. Alberta, Department of Mines and Minerals, *Alberta Oil and Gas Picture, 1947-1962*, and *Annual Reports of the Department of Mines and Minerals*.
2. For a more complete discussion of the role in international taxation of the United States foreign tax credit see, Glenn P. Jenkins, "United States Taxation and the Incentive to Develop Foreign Primary Energy Sources," in *Studies in Energy Tax Policies* (Cambridge, Mass: Ballinger, 1974).
3. Commerce Clearing House, *U.S. Master Tax Guide 1974*, Section 1354, p.464.
4. G.P. Jenkins and Brian D. Wright, "The Taxation of Income of Multinational Corporations: The Case of the U.S. Petroleum Industry," *Review of Economics and Statistics* 57 (1975).

Comment

HARRY F. CAMPBELL

INTRODUCTION

The lack of effective competition in mineral lease markets caused by an unequal distribution of information about mineral prospects amongst a relatively small number of bidders leads to the belief that lease sales will result in a relatively small proportion of the "rent" from natural resource exploitation accruing to the government. Garnaut and Ross argue that the maximization of public revenue from resource exploitation is an objective of public policy; and they propose a Resource Rent Tax (RRT) as a means of maximizing public revenue in the presence of risk and absence of competition in the mineral industry.

Public revenue from the RRT can be expressed as $t.B$, where t is the rate of RRT and B is the tax base. The tax base is the taxable proportion of the total net returns to capital engaged in mineral exploitation in the taxing jurisdiction. The size of the tax base will depend on the size of the capital stock, which is an important determinant of the total returns to capital, and the supply price of investment, which, in the Garnaut-Ross proposal, is a policy parameter determining the taxable proportion of returns to capital. In the long run, the size of the capital stock may be expected to vary directly with the supply price of investment set by the taxing authority and inversely with the rate of RRT and the amount of risk perceived by private investors.

Three operational aspects of the RRT proposal as a means of raising public revenue are examined in this discussion. First, the impact of the RRT scheme on the private investor's perception of risk and, hence, on the size of the capital stock in the long run is examined. Second, the functioning of the RRT scheme in the context of a noncompetitive mineral industry is compared with a publicly operated mineral lease market as a means of raising public revenue. Third, the administrative problems of defining the tax base on a project-by-project basis and preventing tax avoidance at high marginal rates of RRT are considered. The discussion concludes with some comments on the appropriateness of Garnaut and Ross's objective of maximizing public revenue from resource exploitation.

RISK

Garnaut and Ross argue that the principal merit of the RRT scheme is that it does not discourage investment in the mineral industry of the taxing jurisdiction through the imposition of additional risk upon private investors. This claim is supported by the fact that the RRT is collected after the mining project has reached the breakeven point defined by the supply price of investment, and, consequently, that the scheme does not lengthen the expected payback period of the project. Thus, it is implied that the adoption of the RRT proposal would not have the effect of reducing the capital stock and the tax base in the long run.

An analysis of the mining firm's investment decision based on some continuous objective function reflecting both the expected value and the variance of the rate of return defined as a random variable might modify the conclusion that the introduction of the RRT scheme would not affect the flow of investment in the industry. Since the RRT is collected on a project-by-project basis with no loss offset, the scheme has the effect of a leftward shift of the upper tail of the probability distribution of the return to a project with no compensating rightward shift of the lower tail. The result is that, while the variance of the rate of return is reduced, its expected value is also reduced. If the firm's objective function is continuous with respect to the mean and variance of the return, the question of whether a reduction in both mean and variance will affect the decision to invest in a project is an empirical one about the nature of the objective function.

While the Garnaut-Ross proposal may not be completely neutral with respect to a firm's decision to develop a given project, it is probably safe to argue that it imposes less risk on the private investor than a system of auctioning mineral leases. Consequently, it is possible that the RRT scheme would result in a higher annual flow of investment, a larger long-run capital stock, a larger tax base, and hence potentially higher public revenues than a comparable mineral leasing scheme. These potentially higher revenues may be offset to some extent by two costs which the scheme imposes on the public sector. In the first place, the postponement of revenue collection implied by the scheme should be regarded as a cost by a government which has the maximization of the present value of revenues as an objective. Secondly, the reduction in private risk may involve a corresponding increase in public sector risk bearing.

There are two reasons for not regarding a reduction in private risk at the expense of a possible increase in public sector risk as a disadvantage of the RRT scheme. First, it is possible that the increment to the risk of the government's portfolio of assets resulting from the addition of a given project may be less than the increment to the risk of the private firm's portfolio, because of the government's ability to pool the risks associated with mining projects. This argument would rest on the empirical assertion that the government's portfolio contains a significantly larger number of projects than that of a private firm, or that the covariance of the return from an additional mining project and the returns from the existing portfolio of public assets is negative. It should be emphasized that the pooling argument is based on the superior ability of the public sector to absorb risks and not upon a difference in attitude to risk between private and public decision makers; indeed, in this context, the alleged risk aversion of public decison makers is irrelevant, since public risk taking would be institutionalized by the RRT scheme. Second, it is possible that, even in the absence of pooling, the RRT scheme may result in a reduction of private risk without any increase in public risk: the introduction of the RRT proposal or any other viable proposal to collect public revenue might increase investors' confidence in the existing institutional structure and reduce the risk of private losses through transfers of claims on assets from the private to the public sector.

LACK OF COMPETITION

Garnaut and Ross believe that the objective of maximizing public revenue from the taxation of natural resource projects will be achieved by means of a taxing scheme which does not reduce the rate of mineral exploitation in the taxing jurisdiction. They argue that the RRT proposal satisfies this criterion by taxing only those projects which earn a rate of return higher than the supply price of investment and by not adding to the risk in private projects.

The supply price of investment, as defined by Garnaut and Ross, is the private opportunity cost of capital invested in natural resource projects in the taxing jurisdiction. The private opportunity cost of capital can be viewed as the rate of return which could be earned by investing in equally risky natural resource projects in other jurisdictions. If the capital market is imperfect because, for example, information is unequally distributed amongst a few large firms, the private opportunity cost rate of return may be higher than a rate of return representing the social opportunity cost of using capital in the mining industry. The existence of these two concepts of the opportunity cost, or supply price, of capital gives rise to two definitions of *rent*. Garnaut and Ross avoid entering into a controversy regarding the appropriate definition of rent by arguing implicity that a tax base corresponding to the concept of rent as defined by the private opportunity cost of capital will maximize public tax revenue.

It is not clear which, if either, of the two opportunity cost rates is the appropriate choice as a value for the policy parameter Garnaut and Ross refer to as the supply price of investment, given the objective of maximizing tax revenue. The choice of the private opportunity cost rate will result in a larger long-run capital stock (but with a smaller proportion of the returns to capital being included in the tax base) than will the choice of the social rate. The choice of the social opportunity cost rate, on the other hand, will result in a smaller long-run capital stock, but a larger proportion of the returns to capital will be included in the tax base than the choice of the private rate will permit. In other words, it is not clear whether taxing a lower proportion of the returns from a higher capital stock or taxing a higher proportion of the returns from a lower capital stock will yield the most revenue.

If the mining industry is characterized by a lack of effective competition, the public body responsible for administering a RRT scheme may find itself in a position similar to that of a public body responsible for marketing mineral leases. In both cases, the public body may feel impelled to forego some of the rent, as defined by the social opportunity cost of capital, in order to "bribe" investors to continue the flow of investment to the jurisdiction. In the case of a RRT, rent may be foregone by using a high supply price of investment to determine the tax base, while, in the case of mineral leasing, rent may be foregone by accepting low bids on mineral leases. The public body's view of the appropriate size of the bribe may be determined partly by considerations of revenue collection and partly by such other considerations as employment in the jurisdiction.

ADMINISTRATIVE PROBLEMS

As Garnaut and Ross point out, a high rate of RRT would probably encourage tax evasion. Since the RRT scheme involves calculating the tax base on a project-by-project basis, it offers a great deal of scope for understating taxable returns. For example, it may be difficult to evaluate a corporation's allocation of capital costs to specific resource projects or its pricing policies to subsidiaries not subject to RRT. Problems of this kind, which are encountered to a substantial degree by currently employed profits tax schemes applied to resource based multinational corporations, would have to be resolved by balancing marginal tax collections against marginal costs of monitoring tax returns for various possible rates of RRT.

One class of expenditure which is particularly difficult to assign on a project-by-project basis is the cost of exploration. Under the RRT scheme, exploration expenditures attachable to specific projects may be deducted from the revenues of those projects, while general exploratory expenditures which cannot be so attached are nondeductible and may be compensated for by the choice of a higher supply price of investment for all projects. The deductibility of project specific exploration expenditure amounts to a subsidy of this class of expenditure, whereas the higher supply price of investment offers no such immediate incentive to increase general exploration expenditures. It might be argued that it is the latter rather than the former type of exploratory work which should be subsidized on the basis of potential spillover benefits.

CONCLUSION

This discussion has centred on the effectiveness of the RRT proposal as a means of satisfying Garnaut and Ross's objective of maximizing the public revenue from resource exploitation. The appropriateness of this objective, however, depends upon the particular circumstances of the taxing jurisdiction. Garnaut and Ross frame their analysis in the context of a developing economy in which foreign owned firms locate, extract, and export mineral resources. Other types of jurisdictions might be concerned with considerations of consumer and producer surplus, employment creation, and tax neutrality, which may not be recognized by a RRT scheme aimed at revenue maximization.

Discussion of alternative rent collecting schemes tends to be rather barren unless it is tightly focused on a specific set of circumstances. With no empirical information on public and private sector attitudes towards risk, on social and private opportunity costs of capital in the mining industry of the jurisdiction in question, on the elasticity of the schedule of the marginal efficiency of investment in mining, or on administrative costs, the choice between alternative revenue collecting schemes is rather empty.

Investment in Information for the Assessment of Mineral Value: Some Guidelines for Mineral Leasing Policy

BRIAN W. MACKENZIE

The concept of *economic rent* as applied to mineral resources is clear. It is a payment made to the owner of mineral resources to compensate for their use and depletion, the payment being determined by the value of the mineral resources. Economic rent is the *surplus* value over and above all normal cost elements. These costs should include minimum acceptable return on the investment required to find, develop, and produce mineral resources. Surplus value arises because mineral resources are scarce and are variable in quality. There is no doubt that economic rent exists and that it is highly variable among mineral deposits. For example, a study carried out by the author concerning economic discoveries of base metal mineral deposits in Canada showed rates of return varying from 10 per cent to over 60 per cent. The marginal discoveries yielded no economic rent. The higher rates of return reflected the superior quality or grade of particular deposits, indicating a high surplus value-economic rent component.

The provincial government owns the mineral resources within its domain and is entitled to the economic rent. The fundamental issue is how much, in payments, to collect. If too little is collected, part of the economic rent is forfeited as excess profit to the mining company. If too much is collected, there will be an inadequate incentive for future investment. The elusive balance between these factors will result in maximizing discounted provincial government returns in the long term. Given the right level for provincial government returns, the important interrelated issue is how to collect. There are many alternative rent payment systems. Competitive systems may be based on cash bonus, royalty, work commitment, or profit share bidding prior to either the exploration or the development stage of investment. Noncompetitive systems may be based on fixed shares of the revenue, profit, or profitability realized in the course of production. Each system will have a different consequence. As the objective should be not to reduce total provincial returns, the method of collection must be efficient.

The purpose of this article is to suggest an approach for documenting mineral investment experience in a way that provides empirical support for analysis of the mineral leasing policy issues outlined above. The article comprises three interrelated parts. The first describes the stages, information responses, and decision criteria associated with the mineral investment process to provide a framework for the assembly of empirical data. This framework suggests a methodology, outlined in the second part, for assessing the economic characteristics of mineral investment in specific environments of interest, based on empirical data. The third illustrates how such assessments can be applied to provide guidelines for the analysis of mineral leasing policy issues.

THE MINERAL INVESTMENT PROCESS

The mineral investment process is the means of converting minerals from geologic resources into marketable products. The physical occurrence of mineral deposits in nature and the demand for mineral products in the economy provide the basic stimuli. If the relationship between mineral occurrence and market demand is perceived to be favourable, investment activity flows through a number of sequential stages: exploration, development of mineral deposits, and construction of mineral processing facilities. These activities result in the production of mineral products to supply market demand. Changes in demand, depletion, and advances in processing technology reflect the dynamics of the process.

Four types of organization which have direct responsibilities relating to the mineral investment process can be identified: (a) an exploration organization responsible for finding and delineating mineral deposits; (b) a mining organization responsible for developing, producing, and processing mineral resources; (c) a provincial government organization which owns the mineral resources within its domain and assumes the responsibilities of a landlord; (d) a federal-provincial government organization responsible for the development and implementation of mineral policy to ensure that mineral investment activities make the best possible contribution to overall economic and social needs.

The exploration and mining organizations are directly responsible for mineral investment. The provincial and federal-provincial government organizations establish appropriate terms of reference for these investment activities.

Exploration and mining organizations may or may not be integrated within a *mineral resource company*. In any case, it is important to separate these two areas of responsibility to ensure an equitable distribution of returns between them, based on their relative contributions to the investment process.

Determinations of surplus value and economic rent are based on the value of the mineral investment process. Mineral value is determined by the time distribution of cost, risk, and return characteristics through all the stages from primary exploration to mineral production. For a mature mineral investment environment like Canada's, these economic characteristics can be assessed on the basis of historical experience.

For convenience, the mineral investment process may be subdivided into five sequential stages. The fundamental strategic decision to be made by the exploration organization is the *selection of an investment environment* for investment. Thus, the exploration organization makes a commitment to invest within an environment for the *discovery of mineral deposits* and, subsequently, for the *delineation of selected discoveries.* In time, these three information gathering investment stages yield possible economic discoveries for evaluation. The mining organization evaluates each of these development opportunities and decides whether or not each should be developed to production and, if so, how. This is referred to as the *mineral development decision.* The subsequent development and production activities generate increasingly complete information regarding mineral value. The *mineral production process* raises a number of considerations regarding the efficiency of exploitation.

Selection of an Investment Environment

A mineral investment environment comprises the distribution of undiscovered mineral deposits in nature and is defined by geological setting, deposit type, and geographical area; for example: massive sulphide base metal environment in the Canadian Shield; porphyry copper-molybdenum environment in the Canadian Cordillera.

Selection of an investment environment is based on expected return and survival considerations.[1] Expected return assesses the long-term attractiveness of investment to the exploration organization. Survival assesses the organization's short-term difficulties in realizing the expected return due to limited investment funds in relation to environmental risk. Minimum acceptable conditions for investment depend on the objectives of the exploration organization.

The expected return on investment can be assessed in a number of ways. For example, the *expected value of exploration investment* measures the average value that an exploration discovery will yield in the long term, when the successes and failures of a large number of discoveries are considered. For illustrative purposes: $EV = pR-C$, where $EV =$ expected value of a discovery; $p =$ probability of an economic discovery; $R =$ return to the exploration organization given an economic discovery; $C =$ average exploration cost associated with making a discovery.

For example, if $p = 0.01$, $R = $ \$21,000,000, $C = $ \$150,000, then $EV = 0.01(21,000,000)-150,000 = \$60,000$.

If average discovery costs vary significantly among exploration environments, then expected value per discovery does not reflect the attractiveness of exploration investment. In such cases, expected value per unit of exploration costs—say, $100,000—should be used.

The *expected rate of return on exploration investment* measures the discount rate which equates the average investment required to make an economic discovery (C/p) with the return to the exploration organization given an economic discovery, allowing for the time interval between placing the investment and realizing the return.

Expected return criteria assess the long-term favorability of an investment environment. The higher the expected return, the more attractive the environment. Usually an expected value greater than zero and an expected rate of return greater than the cost of capital are regarded as necessary conditions for selection.

An important characteristic of mineral investment environments is the low probability associated with economic discovery. Under this condition, the application of limited organizational funds does not ensure the realization of expected return so exploration resources may be invested without success. This organizational risk introduces a survival element into the selection process, which may be quantified by applying the classical problem of the gambler's ruin.[2]

For example, if $p = 0.01$, $R = \$21,000,000$, and $C = \$150,000$, we can tabulate the survival probabilities for various levels of investment as follows:

Exploration Investment	Probability of Survival
$1,500,000	0.05
$9,000,000	0.27
$15,000,000	0.40
$37,500,000	0.72
$75,000,000	0.92
$150,000,000	0.99

Expected return and survival conditions depend on the inherent geological characteristics of an investment environment, the size, the skills and structure of the exploration organization, and the applicable economic rent payment and taxation systems. The relative importance of the expected return and survival criteria depends on the size and risk preference of the exploration organization. Normally, both factors have an important influence on the selection of an investment environment.

Discovery of Mineral Deposits

Within a selected investment environment, search targets are successively narrowed and the level of detail of information increased by proceeding sequentially through a number of information gathering investment stages. The techniques of applied geology constitute the search method. In general, four sequential stages (see Table 1) may be distinguished prior to discovery: area selection, reconnaissance exploration, follow-up ground exploration, and exploratory drilling. Exploration continues as long as the analysis of information at the end of each stage provides economic justification for further investment. Ultimately, mineral deposits are discovered. Discovery of a mineral deposit may be considered to have been made when exploratory drilling of a target is sufficient to provide the basis for a delineation decision. Thus, a discovery is made on completion of exploratory drilling.

TABLE 1

TYPICAL EXPLORATION PROGRAMS FOR THE DISCOVERY OF
BASE METAL MINERAL DEPOSITS
(CANADIAN SHIELD)

(1)   Selection of 300 square-mile areas
        -   based on geological concepts and available information and experience
(2)   Reconnaissance geophysical survey
        -   airborne electromagnetic survey over selected areas
        -   definition of anomalies
(3)   Follow-up ground geology and geophysics
        -   for each anomaly
        -   property acquisition, line cutting, electromagnetic and magnetic surveys,
             geological mapping
        -   selection of drilling targets based on favourable geophysical responses
             geological settings
(4)   Exploratory drilling
        -   diamond drilling of targets
        -   definition of discoveries
        -   selection of discoveries for delineation

The purpose of sequential investment in the discovery process is to limit the exploration investment associated with making a discovery. The investment per discovery is reduced when situations are perceived to be uneconomic and are rejected before being fully explored. The benefits of the sequential process, namely, the limiting of exploration investment per discovery, must be balanced against the opportunity cost associated with rejecting potential economic discoveries. It is this benefit-to-cost relationship which determines the optimum selectivity exercised in accepting and rejecting opportunities at the end of each sequential stage.

Delineation of Selected Discoveries

Exploratory drilling results in the discovery of mineral deposits. Those discoveries which give positive indications of economic potential are taken to the delineation stage.

Delineation consists of sampling a mineral deposit, normally by diamond drilling, and applying the sampling results to estimate relevant geological characteristics. The objective of the sampling programme is to provide a data base which can be used either to reject the deposit on the evidence provided by partial delineation or to provide the geological estimates required for the economic evaluations associated with the mineral development decision.

Delineation provides estimates of the size and quality of a mineral deposit. Size and quality estimation is a two-step process. Initially, estimates are required for the mineral deposit as a whole. Then, economic conditions (which, for metallic mineral deposits, are expressed as a cut-off grade) are imposed on the overall estimates to assess economic reserves. The classification of economic reserves, traditionally into proven, probable, and possible categories, is based on the level and reliability of the sampling information. Estimates of recovery and dilution factors are used to convert the *in situ* economic reserve assessment to a recoverable basis. The significance of variations in economic conditions on economic reserves depends on the geological features of the mineral deposit, particularly the relationship between size and quality. Knowledge of the size-quality relationship is useful when specific sets of economic conditions are imposed on geological assessments of size and quality to derive estimates of economic reserves.

For metallic mineral deposits, the exponential distribution may be used to estimate the size and quality relationship within a deposit. Using the exponential relationship (discussed fully by the author elsewhere),[3] the size of economic reserves increases geometrically as the grade declines arithmetically. Table 2 presents an exponential size-quality relationship for a hypothetical copper deposit containing 200 million tons with an average grade of 0.5 per cent copper.

The critical decision associated with delineation is choosing the sampling investment which is economically justified. This decision reflects the economic sampling limit. The reliability of size and quality estimates is improved by increasing the sample size, thereby reducing the estimated standard error of the mean and the spread of confidence limits about the mean value estimate. However, the benefits from increased sampling are subject to diminishing returns; that is, as sample size is increased, the standard error of the mean is reduced, but at a decreasing rate. At some point, marginal benefits from sampling investment just balance marginal costs. This point defines the economic sampling limit.

TABLE 2

Cut-Off Grade (% Cu)	Ore Reserves (tons)	Metal Content (tons)	Average Grade (% Cu)
0.00	200,000,000	1,000,000	0.50
0.50	73,580,000	735,800	1.00
1.00	27,070,000	406,000	1.50
1.50	9,960,000	199,100	2.00
2.00	3,660,000	91,600	2.50
2.50	1,350,000	40,400	3.00
3.00	496,000	17,350	3.50
3.50	182,000	7,290	4.00
4.00	67,000	3,020	4.50

On completion of delineation, the mineral deposit is transferred from the exploration organization to the mining organization. While the specific terms of this transfer vary, they are reflected in R, the return to the exploration organization given an economic discovery.

Mineral Development Decision[4]

The mineral development decision, taken by the mining organization on completion of delineation, determines whether or not a mineral deposit should be developed to production and, if so, the optimum mineral development parameters. This involves the selection of an optimum mineral development alternative from a number of technically feasible alternatives followed by the comparison of this optimum with those of other available investment opportunities. If the optimum is sufficiently attractive, the deposit is developed to production. The selection process is carried out within a framework of organizational objectives and resources.

The investment decision is based on the optimization of mineral development parameters, which include capacity and, for metallic mineral deposits, the cut-off grade. On the basis of available information and experience, revenue and cost estimates are made for each feasible combination of development parameters. Economic evaluation techniques are applied to these estimates to reduce each alternative to economic criteria which will support the investment decision.

Economic evaluation of investment alternatives is based on the limited information available concerning economic reserves, costs, and mineral markets. To compare and select alternatives, both the expected value and the uncertainty of profitability should be evaluated. In addition to information which can be quantified in terms of expected profitability and uncertainty, intangible factors must also be considered in the decision process. Possibilities for extending economic reserves beyond those delineated at the time of the evaluation may constitute an important intangible.

Cash flow and time value are the basic concepts used to assess profitability. *Net present value* may be used to compare the profitability characteristics of alternatives, given a fixed level of investment (fixed capacity, for example). To estimate net present value, positive and negative annual cash flows are discounted at a predetermined interest rate representing the cost of capital, and the discounted values are summed. The *rate of return*, one of the most widely used profitability criteria, is the discount rate which equates the present value of negative cash flows with the present value of positive cash flows, that is, the discount rate which produces a zero present value.

The economic evaluation of feasible mineral development alternatives with respect to both expected profitability and uncertainty provides the basis for optimizing mineral development. The mining organization, motivated by the desire for profitable investment, prefers a development alternative with a higher expected profitability to one with less. At the same time, it is averse to uncertainty and, therefore, prefers a development alternative with less uncertainty to one with more. Uncertainty aversion reflects a concern with ensuring some minimum level of profitability. Thus, while uncertainty is assessed by the overall probability distribution of the profitability outcome, its critical component is the lower confidence limit. In economic terms, the organization will select the alternative which maximizes its "utility."

Mineral Production Process

For the mineral operation, investment is sunk in exploration and mineral development, and decisions are of a tactical nature. Operational decisions embody such elements as cost control, production planning, productivity, the development and implementation of new production and processing technology, and equipment replacement. The effectiveness of these policies will be reflected in cost trends which, in turn, will determine changes in economic reserves and productive life. A primary objective of the mining organization with respect to a mineral operation is the maximization of total profit over mine life. This objective will be achieved by maximizing economic reserves and productive life and by mining to the extensive economic margin. For metallic mineral deposits, the cut-off grade defines economic reserves and productive life. The cut-off grade is the minimum grade which can be economically mined. For operational conditions where the objective is maximization of total profit over mine life, cut-off grade is determined by the equation of marginal cost and marginal revenue. At the margin, grade is just sufficient to offset cost: $MC = MR = PGR/(1 + D)$, or $G = MC(1 + D)/PR$, where MC = cost of mining and processing a marginal ton of ore (\$); MR = revenue realized from a marginal ton or ore (\$); P = net price of metal recovered (\$ per ton); G = cut-off grade, percentage content of mineral product in marginal ton of ore; R = mill recovery factor (metal recovered/metal content of ore reserves); and D = dilution factor (waste mined/ore mined).

Thus, cut-off grade varies directly with the cost of producing a saleable mineral product and inversely with the price received for the product. The significance of variations in cut-off grade for operational decisions depends on the geological parameters of the deposit, particularly the previously discussed relationship between tonnage and grade.

The determination of cut-off grade and economic reserves for a metal mining operation may be strongly influenced by government policy considerations. Mineral conservation policy may require mining to the average grade of ore reserves as defined by the economic margin. The structure of economic rent and taxation payments may raise the cut-off grade.

ASSESSMENT OF ECONOMIC CHARACTERISTICS FOR AN INVESTMENT ENVIRONMENT

The first part of this article described the stages, information responses, and decision criteria associated with the mineral investment process. This background provides a basis for assessing the economic characteristics of mineral investment in specific environments of interest, and a methodology for making such assessments will now be discussed. What is the relevance of these assessments for the analysis of a mineral leasing policy?

Mineral leasing policy issues include determination of economic rent and evaluation of the relationship between possible levels and structures of a rent payment system and the investment behaviour of exploration and mining organizations. The empirical assessment of economic characteristics suggested here provides practical assistance in analyzing these issues. From the definition of economic rent and the description of the mineral investment process, it is clear that three economic characteristics are of fundamental importance in the analysis of these issues: the average exploration cost associated with making a discovery; the probability of discovering an economic mineral deposit; the return resulting from an economic discovery. The definition and measurement of these parameters present many practical difficulties. In general, mineral exploration is characterized by a very low probability of economic discovery, and a very large return given an economic discovery relative to the exploration cost associated with discovering a deposit. For a mature investment environment like Canada's with its adequate history of mineral investment activity, these characteristics can be assessed using documented empirical data on past trends in investment experience.

At the outset, it is important to note two considerations embodied in the suggested methodology. First, the empirical data base is assembled for an investment environment as a whole in order to have a meaningful sample size of experience for analysis. Thus, assessments of economic characteristics based on these data reflect the *average* performance of investment. Variations in performance among participating organizations, which occur for reasons described previously, are not considered. Secondly, the empirical data and assessments described herein are *exclusive* of provincial rent payment considerations. For example, the *return resulting from an economic discovery* is assessed as a total return net of federal tax. Thus, the assessed return consists of the total amount which is to be shared among the exploration organization, the mining organization, and the provincial government. The reason for excluding rent payments from these assessments is to provide a data base which may be generally applied to determine economic rent and to study the effects of alternative rent payment systems on the economic characteristics of investment, and, thereby, on the investment behaviour of exploration and mining organizations.

Average Discovery Cost

The average exploration cost of a discovery is assessed on the basis of the stages, exploration techniques, costs, and information responses associated with typical exploration programmes for the deposit type and region considered. The estimation of average discovery cost is based on discussions with exploration personnel directly experienced in managing exploration programmes.

Assessment of average discovery cost requires a definition of *discovery* which must be consistently applied. Normally, discovery is considered to be associated with completion of the exploratory drilling prior to delineation. For example, a discovery may be defined when exploratory drilling is sufficient to provide the basis for a delineation drilling decision. Thus, a discovery is made on completion of the last stage of exploratory drilling.

To illustrate the assessment of average discovery cost, consider the following typical exploration programme for copper-zinc deposits in the Canadian Shield. Initially, favourable 300-square-mile areas are selected. Area selection is conceptually based. The cost of research for selecting an area is two months of a regional geologist's time at $24,000 per year. The areas selected are subjected to an airborne electromagnetic survey using quarter-mile spacing. Airborne survey costs average $40 per line mile. After airborne work, a variable number of anomalies noted in the airborne survey are selected for investigation on the ground. On the average, 60 anomalies are selected. Follow-up ground exploration includes property acquisition, line cutting, electromagnetic and magnetic surveying, and geological mapping with costs averaging $11,000 per square mile. Follow-up grids average 0.2 square miles. On the average, one-third of the follow-up anomalies show favourable geophysical responses and geological settings, and these are drilled. Of the anomalies drilled, 5 per cent justify an additional four holes of exploratory drilling. These represent discoveries. Exploratory drilling averages 300 feet per hole and costs $20 per foot. Twenty-five per cent of these discoveries warrant delineation. The amount of delineation drilling required varies with the size and variability of the deposit. However, a typical delineation programme results in the proving-up of 200,000 tons, requiring one foot of drilling for the establishment of twenty-five tons of reserves at a unit cost of $15 per foot. Cost summary for an area may be tabulated as follows:

Area selection: 24,000/6 =	$4,000
Reconnaissance survey: 1200(40) =	$48,000
Ground follow-up: 60(0.2) (11,000) =	$132,000
Exploratory drilling: 20(300) (20) + 4(300) (20) =	$144,000
Delineation drilling: 0.25(8,000) (15) =	$30,000
Total exploration cost:	$358,000
Number of deposits discovered:	1
Average exploration cost per discovery:	$358,000

If exploration cost is an allowable deduction for federal tax purposes, and to the extent that income is available against which deductions may be written off, the before tax exploration cost estimated above may be significantly reduced on an after tax basis. It is this real after tax cost to the exploration organization which should be applied when assessing the economic characteristics of exploration investment. For example, if the effective federal mining taxation rate is 30 per cent, the average exploration cost per discovery in the foregoing example would be reduced to 0.7 (358,000) = $250,600.

Probability of Economic Discovery

The probability of economic discovery and the potential economic return parameters are more difficult to assess because the economic characteristics of undiscovered deposits are unknown. However, for exploration environments with a reasonable history of exploration, these parameters can be estimated on the basis of past exploration results. Thus, the probability of discovering an economic mineral deposit is estimated by the historical trend in the ratio of economic discoveries to total discoveries. It is not usually possible to measure directly the total number of discoveries made, but this parameter can be derived from estimates of primary exploration expenditure and the average exploration cost of making a discovery.

An *economic discovery* is defined by minimum acceptable size and profitability criteria, under conditions of present-day technological and economic conditions. The minimum size condition ensures that mine development opportunities are sufficiently large to make a significant contribution to overall organizational performance. The minimum profitability condition will be determined by the mining organization's cost of capital. For this purpose, size may be assessed according to the total revenue which a deposit generates, and profitability by the indicated rate of return.

Return from Economic Discovery

The total return resulting from an economic discovery is determined on the basis of the size and profitability characteristics of past economic discoveries assuming their development under present-day conditions. Initially, possible economic discoveries are listed for evaluation. Revenue and cost characteristics for the evaluation of possible economic discoveries are estimated on the basis of current technological and economic conditions. General estimates of metal prices and smelter payments, together with specific ore reserve, recovery and dilution factors, and mill capacity inputs are used to calculate annual revenue and total revenue over the mine life. Mill capacity and unit operating costs are used to estimate annual operating cost. The resulting income before tax is subjected to federal taxation. Annual cash flows are then calculated as the difference between revenue and the operating, capital, and taxation costs. The annual cash flow estimates are used to calculate rate of return and net present value. Results are compared with minimum acceptable size and profitability conditions to define economic discoveries.

Specific estimates of ore reserves, mine recovery and dilution, mill recovery, mill capacity, capital costs, preproduction development time and operating costs are made for the evaluation of each possible economic discovery. These estimates are based on published data and on information obtained from mining organizations and government agencies. In cases where specific data are not available, estimates are made on the basis of generalized relationships reflecting current knowledge of the evaluation, development, and operation of mines in the investment environment under study.

The method of evaluating possible economic discoveries, as outlined above, does not aim at simulating the actual profitabilities realized. The deposits considered will have been discovered, developed, and operated under widely different economic conditions. To evaluate the characteristics of economic discoveries helpful to the present-day formulation of mineral leasing policy requires imposing a common set of conditions on each economic discovery, namely, current technological and economic conditions.

Hypothetical Example

The following simplified example is an assessment of economic characteristics for a hypothetical base metal environment. The results are used to show how such assessments can be applied as guidelines for mineral leasing policy.

Average discovery cost. The average discovery cost is assessed on the basis of a typical exploration programme which reflects current thinking and experience regarding the most efficient exploration approach within the hypothetical base metal environment. A discovery is defined as completion of the last exploratory drilling stage.

The average discovery cost is $500,000. The average discovery cost *to the exploration organization* is reduced if exploration expenditure is an allowable deduction for federal tax purposes and income is available against which deductions can be written off. Assuming an effective federal tax rate of 30 per cent, the average after tax discovery cost to the exploration organization is $350,000.

Return resulting from an economic discovery. This return is assessed net of federal taxation considerations for discoveries made during the period 1960-73. Consideration is given as to the sharing of this return by the exploration organization, the mining organization, and the provincial government.

An economic discovery is defined by minimum acceptable size and profitability criteria *to the mining organization* under conditions of

FIGURE 1

HYPOTHETICAL BASE METAL ENVIRONMENT TOTAL RETURN RESULTING FROM AN ECONOMIC DISCOVERY

Revenue (mill. $)

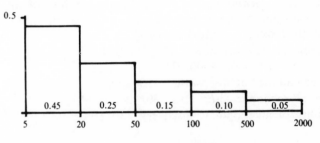

Cash Flow (mill. $)

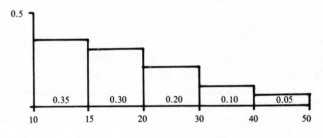

Rate of Return (%)

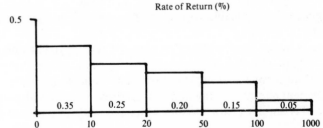

Net Present Value at 10 per cent (mill. $)

present-day technology and economics. With respect to size, it is assumed that a deposit must generate a total revenue of at least $20 million. With respect to profitability, it is assumed that a deposit must yield a 10 per cent rate of return to be classified as an economic discovery.

Frequency distributions of revenue, cash flow, rate of return, and net present value at 10 per cent (discounted to the start of mineral development) are shown in Figure 1. The expected return characteristics for an economic discovery are as follows:

Revenue: $250 million
Cash flow: $75 million
Rate of return: 18 per cent
Net present value at 10 per cent: $45 million

Probability of an economic discovery. The probability of an economic discovery is estimated on the basis of exploration results within the environment for the 1960-73 periods as follows:

Total exploration expenditure: $750,000,000
Average discovery cost: $500,000
Estimated number of discoveries: 1500
Number of economic discoveries: 30*
Probability of an economic discovery = 30/1500 = 0.02

SOME GUIDELINES FOR MINERAL LEASING POLICY

The purpose of assessing the economic characteristics of mineral investment as suggested above is to provide an empirical data base for determining rentability and evaluating the effects of alternative rent payment systems on mineral investment conditions within a particular environment. Thus, guidelines may be derived for determining the most efficient rent payment system to be used in practice.

The probability, cost, and return characteristics of mineral investment are assessed exclusive of rent payment considerations. This is to provide a data base generally applicable to the assessment of rentability and rent payment effects.

There is a complex interrelationship between rent payment alternatives and the three economic characteristics which determine mineral investment behaviour within a particular environment. This concluding section shows how rent payment alternatives are imposed on these characteristics to derive assessments of economic characteristics which *include* rent payment effects.

---

*Based on the evaluation of economic discoveries outlined above and the estimation of actual discovery dates for each.

These assessments are applied to determine economic rent and guide the selection of a rent payment system.

How does a rent payment system affect the economic characteristics of mineral investment? The average discovery cost to the exploration organization is only affected to the extent that exploration expenditures are allowed as a deduction in computing rent payment liability. The example which follows assumes that exploration expenditures are *not* an allowable deduction for rent payment purposes. The probability of economic discovery is estimated by the historical trend in the ratio of economic discoveries to total discoveries. An economic discovery is defined by minimum acceptable size and profitability criteria. The number of economic discoveries is a function of the rent payment system, if this payment is levied at the investment margin. The higher the incidence of rent payment at the margin, the lower the number of economic discoveries, and, consequently, the lower the probability of economic discovery. The return resulting from an economic discovery is determined by the distribution of economic discoveries above the investment margin. The higher the incidence of rent payment at the margin, the lower the number of marginal economic discoveries included in this distribution and, consequently, the higher the average return characteristic of the distribution.

In evaluating the effects of rent payment alternatives on these economic characteristics, it is convenient to make a distinction between the *level* and the *structure* of rent payment. The level of rent payment refers to the rate of payment and is defined here as *the proportion of the balance of the return from an economic discovery which, after deduction of minimum acceptable returns on investment to exploration and mining organizations, is collected by the provincial government.* The structure of rent payment refers to the manner in which this payment is collected.

The hypothetical example described earlier is used here to illustrate the effects of different levels and structures of rent payment on economic characteristics in order to assess resulting changes in investment conditions and economic rent.

Hypothetical Example

The assessment of economic characteristics for the hypothetical example is summarized as follows:

a. Average discovery cost: after-tax discovery cost to the exploration organization, — $350,000
b. Probability of economic discovery: exclusive of rent payment considerations, — 0.02.
c. Return resulting from an economic discovery: average return after federal tax, expressed as net present value at 10 per cent, —$45 million.

Three levels of rent payment are considered, expressed in each case as a proportion of the balance of return after deducting all costs:

Level 1—90 per cent
Level 2—70 per cent
Level 3—50 per cent

For each level of rent payment, three structures of payment are considered:

Revenue   — Percentage of revenue: payment not directly related to the investment margin for economic discovery.
Profit   — Percentage of profit: payment reduced at the investment margin, vanishing as cost equals revenue.
Profitability   — Percentage of profit after realization of minimum acceptable rate of return: payment reduced to zero as the investment margin for economic discovery is approached.

What are the effects of each level-structure combination on the basic economic characteristics? There is no effect on average discovery cost. The effects on probability and return characteristics depend on both the level and structure of rent payment. The profitability structure does not affect these characteristics because it does not apply at the investment margin. The revenue and profit forms of payment are levied at the investment margin, thereby reducing the number of economic discoveries and the probability of economic discovery and increasing the average return characteristic. The effects of the revenue form of payment will be more pronounced, as the incidence of payment is independent of the investment margin.

Table 3 shows the effects on the probability of economic discovery characteristic for the second level of taxation. Obviously, the effects of the profit and revenue payment structures become more pronounced as the level of payment increases. Table 4 gives the complete set of effects.

Motivation of Exploration Investment

The investment behaviour of the exploration organization is determined by the expected return and survival conditions described previously. These conditions are based on the probability, cost, and return characteristics of particular investment environments.

The survival criterion implies that the exploration organization is risk averse. Securing adequate survival conditions is an important aspect of exploration planning. The widespread use of joint ventures in mineral exploration indicates the importance of survival considerations. However, the fundamental motivation of the exploration organization is as a risk

TABLE 3

PROBABILITY OF ECONOMIC DISCOVERY – LEVEL 2

Payment Structure	No. of Discoveries 1960-73	No. of Economic Discoveries 1960-73	Probability of an Economic Discovery
Revenue	1500	18	0.012
Profit	1500	24	0.016
Profitability	1500	30	0.020

TABLE 4

EFFECT OF RENT PAYMENT ALTERNATIVES
ON PROBABILITY AND RETURN CHARACTERISTICS

Payment Structure	Economic Characteristic	Level 1	Level 2	Level 3
Revenue	Probability	0.008	0.012	0.016
	Return (x $1,000,000)	55.0	50.0	47.0
Profit	Probability	0.014	0.016	0.018
	Return (x $1,000,000)	50.0	47.5	46.0
Profitability	Probability	0.020	0.020	0.020
	Return (x $1,000,000)	45.0	45.0	45.0
	Avg. Discovery cost ($)	350,000		

taker—the motivation is to invest relatively modest discovery costs for a small chance of a spectacular return.

The hypothetical example illustrates the effects of alternative rent payment systems on the economic characteristics which guide exploration investment. As these characteristics are changed by the rent payment system within an environment, so is the investment behaviour of the exploration organization.

The economic characteristics of mineral investment are fixed for a particular rent payment system, and the motivation of the exploration organization to invest is determined by the share of the return from an economic discovery which it is able to realize. This return is not directly recovered by the exploration organization. It is part of the return realized by the mining organization from subsequent mineral production. How large should this share be to provide an adequate motivation to invest?

The exploration organization has an essential role in the mineral investment process and its investment effort should yield an equitable return. In the long term the minimum acceptable rate of return required to motivate exploration investment within an environment is the cost of exploration capital employed. To motivate exploration investment, this rate of return should not be ensurable to individual organizations; rather, it should be the average rate of return for exploration investment within an environment. In this way, motivation would be towards doing better than average. To ensure such a breakeven return might be sufficient to compensate for short-term survival costs.

The economic characteristics assessed reflect the average performance of exploration investment within an environment. There are obvious benefits for an exploration organization which applies superior skills and selectivity in exploration; benefits in terms of increasing the probability of an economic discovery, lowering the average discovery cost, and improving the return characteristics of economic discoveries. In an exploration environment which is characterized by a minimum acceptable average rate of return but a high level of risk, the selection of exploration opportunities of exceptional merit will be a necessary condition for the survival of individual exploration organizations.[5] This will result from the application of superior skills, developed within an exploration organization over time through persistent exposure to a particular exploration environment. Thus, a minimum acceptable average rate of return will tend to select the better than average exploration organizations for survival and long-term prosperity; other organizations will with time be ruined.

The assessment of economic characteristics provides guidelines for determining the share of the return from an economic discovery which an exploration organization should receive. This is illustrated by the hypothetical example. Assume that the minimum acceptable average rate of return within the hypothetical base metal environment is deemed to be 10 per cent and that there is a two-year interval between exploration expenditure and the start of mine development. Then, considering the profitability form of rent payment:

average cost to the exploration organization per economic discovery = $350,000/0.02 = \$17,500,000$, and
accumulated value of expenditure at the start of mine development = $17,500,000 (1.10)^2 = \$21,200,000.$

Thus, the exploration organization making an economic discovery should receive \$21.2 million of the \$45 million average net present value or, generally, about 47 per cent of the net present value of an economic discovery in order to recover all exploration costs and provide a long-term incentive for exploration investment.

The minimum acceptable returns required for each of the nine rent payment alternatives are given in Table 5.

TABLE 5

MINIMUM ACCEPTABLE RETURNS TO THE EXPLORATION ORGANIZATION

		Level 1	Level 2	Level 3
Revenue:	Average Cost per Economic Discovery (x$1,000,000)	43.8	29.2	21.9
	Return at Start of Development (x$1,000,000)	52.9	35.3	26.5
	Share of Return (%)	96.0	71.0	56.0
Profit:	Average Cost per Economic Discovery (x$1,000,000)	25.0	21.9	19.4
	Return at Start of Development (x$1,000,000)	30.2	26.5	23.5
	Share of Return (%)	61.0	56.0	51.0
Profitability:	Average Cost per Economic Discovery (x$1,000,000)	17.5	17.5	17.5
	Return at Start of Development (x$1,000,000)	21.2	21.2	21.2
	Share of Return (%)	47.0	47.0	47.0

The return to the exploration organization could take the form of a lump sum payment prior to development or, more realistically, a carried interest through the development and operating stages.

Motivation of Mineral Development Investment

The mining organization requires a minimum acceptable rate of return on investment in mineral development to recover its cost of capital. As discussed previously, the investment behaviour of the mining organization is determined by the relationship between the distribution of possible rates of return for an investment (particularly the expected value and lower limit of the distribution) and its minimum acceptable rate of return. A rent payment system affects investment behaviour by shifting the distribution of possible rates of return in relation to the investment margin.

Economic discoveries, by definition, realize a return above the investment margin. This represents the return from an economic discovery and is reflected by net present value, discounted at the cost of capital rate to the start of mineral development. As discussed in the foregoing section, a share of this return should be allocated to the exploration organization which has made the economic discovery to provide a minimum acceptable return on exploration investment. The balance should be shared between the mining organization and the provincial government.

The rationale for the mining organization's share is associated with the management and technical skills it provides for mine development and operation. Its share of the balance of the return is required to encourage efficient mineral development and operation, innovation, and advances in technology. This payment constitutes a "management fee."

Table 6 outlines the return to the mining organization above the mine development investment margin for the rent payment alternatives considered in the hypothetical base metal sample.

TABLE 6

RETURN TO THE MINING ORGANIZATION ABOVE INVESTMENT MARGIN
(x$1,000,000)

		Level 1	Level 2	Level 3
Revenue:	Balance of Return*	2.1	19.8	23.8
	Return to Mining Organization	0.2	2.0	2.4
Profit:	Balance of Return*	14.7	21.0	23.8
	Return to Mining Organization	4.4	6.3	7.2
Profitability:	Balance of Return*	20.5	22.5	23.8
	Return to Mining Organization	10.3	11.2	11.9

* Net of minimum acceptable return to the exploration organization.

Rentability of Mineral Resources

The provincial government owns the mineral resources within its domain and, thus, is entitled to the economic rent. The economic rent is a surplus value which is realized after minimum acceptable returns have been allocated to the exploration and mining organizations for their investment and management contributions. Suggestions have been made above concerning the minimum acceptable returns required to motivate investment by the exploration and mining organizations and to encourage efficient development and operating practices. Considerable uncertainty is associated with these judgments.

The objective of the provincial government as landlord is to maximize provincial revenues from mineral development. This will be achieved in the long term by maximizing the *expected rent payment from a discovery*.[6] The expected rent payment from a discovery is the product of the probability of economic discovery and the *expected rent payment from an economic discovery*. The expected rent payment from an economic discovery is the expected return minus returns to the exploration and mining organizations as assessed above. The expected rent payment (ERP) from an economic discovery and from a discovery are derived in Table 7 for the rent payment alternatives assumed in the hypothetical base metal environment.

TABLE 7

EXPECTED RENT PAYMENT (ERP)
(x$1,000)

			Level 1	Level 2	Level 3
Revenue:	ERP:	Economic Discovery	1,850	10,290	10,270
		Discovery	15	123	164
Profit:	ERP:	Economic Discovery	17,770	14,730	11,240
		Discovery	249	236	202
Profitability:	ERP:	Economic Discovery	21,450	16,680	11,910
		Discovery	429	334	238

The above results are based on hypothetical data. Nevertheless, a number of interesting points are raised which should be considered in practice.

a. For the revenue structure, provincial government revenue is maximized at the lowest level of rent payment.
b. The profitability structure yields higher provincial government revenues than either the profit or the revenue structures.
c. The highest level of rent payment coupled with the profitability structure yields the highest provincial government revenue. This does not consider the costs associated with inadequate motivation for development and operating efficiency. Consequently, in practice, a lower level of rent payment, for example, Level 2, may yield the best results.

The return to the mining organization above the investment margin may be considered as payment for a fixed contribution. Thus, its share of the balance of the return would be an inverse function of the rate of return on investment in mine development. This is consistent with the concept of economic rent which arises because of the superior quality of particular mineral deposits. A relatively high rate of return on mineral development normally reflects the superior quality of the mineral deposit itself. Thus, the higher the rate of return on mineral development, the greater the proportion of the balance of return which should be allocated to the provincial government.

These factors suggest that shares in the balance of return of the mining organization and the provincial government should be placed on a sliding scale as a function of rate of return to reflect their relative contributions more realistically. This type of payment system is shown in Table 8 for Level 2 conditions.

The hypothetical example shows that both the level and structure of the rent payment system significantly affect the investment behaviour of exploration and mining organizations and the long-term provincial

TABLE 8

SLIDING SCALE RENT PAYMENT – LEVEL 2

Rate of Return (%)	Frequency of Occurrence	Share of Balance of Return	
		Provincial Government (%)	Mining Organization (%)
10–15	0.35	55	45
15–30	0.50	75	25
30–50	0.15	85	15
Average: 18	1.00	Average: 70	Average: 30

government revenues which will be realized from mineral development. The suggested methodology provides guidelines with respect to the optimum level of rent payment. The profitability structure of rent payment has apparent advantages over the other types of structure assessed.

There are two forms of rent payment system which could be based on the concept of profitability:

a. A noncompetitive allocation system whereby the terms of the profitability system are negotiated or, preferably, legislated. These fixed terms are applied to the actual profitability which is realized in the course of production.
b. A competitive allocation system whereby some of the terms of the profitability system, the base rate and the sharing of returns above the base rate, for example, are determined by competitive bidding. The bidding is based on the information available at the time of bidding.

The most efficient form of profitability system to apply in practice would have to be fully analyzed. The objective should be to combine the most desirable features of the noncompetitive and competitive forms.

Motivation of Efficient Mineral Production

An important responsibility of the provincial landlord should be to ensure the efficient operation of mineral deposits. Operating efficiency will be achieved by maximizing mineral recovery, economic reserves, and productive life. For base metal deposits, the cut-off grade defines these parameters.

The determination of cut-off grade for a mining operation may be strongly influenced by mineral leasing policy. In some cases, the terms of mineral leasing agreements require mining to the average grade of economic reserves. The structure of the rent payment system may act to raise cut-off grade and reduce economic reserves. The following hypothetical example compares the relative effects of profit and revenue structures of rent payment.

A copper deposit is assumed, containing 200 million tons at an average grade of 0.5 per cent copper. The tonnage-grade relationship relationship is exponential. A mine capacity of 3 million tons of ore per year has been installed at a capital cost of $37.5 million. Other operating parameters are as follows:

Operating cost: $3.60 per ton of ore mined
Recovery factor: 80 per cent
Dilution: nil
Price: $750 per ton of copper recovered in concentrate
Level of lease payment: $50 million

The example compares the optimization of the mining operation for two structures of rent payment.

Profit: t_p,   percentage of taxable profit; operating and capital costs are allowable deductions.

Revenue: t_r, percentage of revenue.

It is assumed that the operating objective of the mining organization is to maximize total profit over mine life. Time value is neglected.

a. *Profit Payment*
  $G = MC\ (1 + D)/PR = 3.60/750(0.8) = 0.60$ per cent copper
  Applying the exponential tonnage-to-grade distribution:
  Ore reserves = 60,239,000 tons
  Copper content = 662,600 tons
  Average grade of ore reserves = 1.10 per cent copper

Total Revenue = 662,600 (0.8)(750) =	$397,560,000
Operating Cost = 60,239,000 (3.60) =	216,860,000
Operating Profit	180,718,000
Capital Cost	37,500,000
Profit for Rent Payment	143,218,000
Rent Payment	50,000,000
Total Profit (Cash Flow)	$ 93,218,000

Rent Payment rate $= t_p = 50,000,000/143,218,000 = 34.9$ per cent

Mine life $= 60,239,000/3,000,000 = 20.1$ years

b. *Revenue Payment*

$$G = [MC + t_r(PRG)]/PR = 0.006/(1 - t_r) \tag{1}$$

$$T = t_r \text{ (copper content of ore reserves) } (PR) = \$50,000,000$$
$$\text{copper content of ore reserves } = 50,000,000/600t_r \tag{2}$$

Solutions for $G t_r$ and the copper content of ore reserves are obtained by solving expressions (1) and (2) simultaneously with the exponential tonnage-to-grade relationship:

Cut-off grade $= 0.70$ per cent copper
Ore reserves $= 43,319,000$ tons
Copper content $= 591,800$ tons
Average grade of ore reserves $= 1.20$ per cent copper

Total revenue $= 591,800(0.8)(750) =$	\$355,100,000
Rent Payment	50,000,000
Operating cost $= 49,319,000(3.60) =$	177,550,000
Capital cost	37,500,000
Total profit (cash flow)	90,050,000

Rent payment rate $= t_r = 50,000,000/355,100,000 = 14.2$ per cent
Mine life $= 49,319,000/3,000,000 = 16.4$ years

Effects of the structure of rent payment on the efficiency of mineral production, as presented above, should guide the development of a mineral leasing policy. The detrimental effects of the revenue structure of rent payment in this respect are clearly illustrated.

CONCLUSION

The development of mineral leasing policy requires a practical understanding of the mineral investment process through exploration, development and production stages of activity. Within this process decisions are required concerning the selection of investment environments, the discovery of mineral deposits, delineation drilling investment, the development of productive capacity, and the efficiency of mineral production. Expected return and risk criteria are evaluated to support these decisions based on the information which is available at particular decision points.

An understanding of the mineral investment process provides the basis for assessing the economic characteristics of mineral investment in specific environments of interest. Three economic characteristics are of fundamental importance in the long term—average discovery cost, probability of economic discovery, and return from economic discovery. In mature mineral investment environments these characteristics may be assessed on the basis of empirical data and applied to evaluate expected return and risk criteria.

The assessment of economic characteristics provides the information required to analyze how alternative levels and structures of government rent payment affect the motivation of exploration investment, mineral development investment, and the efficiency of mineral production. At the same time, the assessment of economic characteristics enables judgments to be made concerning the rentability of mineral resources and the level and structure of rent payment required to maximize government revenues from mineral resources. This information is important in the formulation of mineral leasing policy.

Notes

1. Brian W. Mackenzie, "Corporate Exploration Strategies," in *Application of Computer Methods in the Mineral Industry* (Johannesburg: SAIMM, 1973), pp. 1-8.

2. This problem is fully described in Brian W. Mackenzie, "A Decision-Making Theory of the Mining Firm," in *Proceedings of the Council of Economics* (New York:AIME, 1969), pp. 321-43, and Mackenzie, "Corporate Exploration Strategies."

3. Brian W. Mackenzie and J.E.G. Schwellnus, "Tonnage-Grade Estimation for Mineral Deposits and the Assessment of Ore Reserves." (Paper delivered at United Nations Inter-Regional Seminar on Advanced Mining Technology, Ottawa, 1973).

4. The analysis framework summarized here is based on more detailed treatments of the subject in Brian W. Mackenzie, "Evaluating the Economics of Mine Development," *Canadian Mining Journal* 91 (December 1970):43-47, and 92(March 1971):46-54, and in Brian W. Mackenzie, M. Bilodeau, and G. Mascall, "The Effect of Uncertainty on the Optimization of Mine Development," *Proceedings of the Twelfth Symposium on the Application of Computers and Mathematics in the Minerals Industry* (Golden, Colorado, 1974), pp. A8-A38.

5. The high level of uncertainty precludes the exploration organization with limited funds from averaging out.

6. This maximizes the provincial government revenue per unit of primary exploration investment if the average discovery cost is fixed (for example, $350,000 in the hypothetical case). The long-term incentive for the exploration organization to invest is the same for each rent payment alternative, namely, an expected return of zero in the hypothetical example. Thus, the long-term level of primary investment will be the same for each alternative. So, the rent payment alternative offering the highest *expected rent payment from a discovery* will maximize provincial revenues.

PART TWO

Problems of Leasing Policy

The Problem of Timing in Resource Development

G.D. QUIRIN
B.A. KALYMON

Our singular title masks a multiplicity of timing problems associated with the development of natural resources. Although they are conceptually distinct, a unity is imposed by the need to produce a single decision for a single development project which is either optimal respecting the several aspects of the problem or satisfactorily resolves conflicts between them. The distinct problems can be identified as follows:

a. the *scheduling problem* of selecting an optimum producing rate with respect to an investment made at a given point in time;

b. the *historical timing problem* of selecting an appropriate time for undertaking a given project considered in isolation;

c. the *cyclical timing problem* of selecting investment priorities between competing projects which might, on grounds of historical timing alone, all be undertaken simultaneously, but which cannot be carried out all at once because of constraints arising out of manpower availability, environmental impact, capital market weakness, balance of payments problems, or similar considerations.

Consideration of timing problems has occupied a prominent place in the literature of resource economics. Much of this previous discussion, however, has been focused on the scheduling problem and, to a lesser extent, on historical timing, while cyclical timing has been comparatively neglected. Recent discussions in Canada, however, arising out of proposals to undertake major projects involving individual expenditures from $1 billion to $6 billion, which are large in relation to normal levels of capital investment in the economy, suggest cyclical timing is a problem of some importance. The number of tar sands recovery plants which we will be able to build at any one time is not known, but it is likely to be relatively small. Nor is it evident that we could simultaneously complete two gas piplines from different parts of the Arctic, an oil pipline from the Mackenzie delta, and the James Bay power project, even if we reject as without foundation the arguments of those who urge that none of these ventures be undertaken.

In this article we seek to develop optimality criteria for timing decisions, to examine the institutional structure within which they are made, and to examine how the role of government, which is the basic owner of the underlying resources, might be modified to improve this aspect of resource utilization.

Our primary attention is focused on problems of historical and cyclical timing. Scheduling of output has been adequately dealt with elsewhere by ourselves and others, [1] frequently in a context which considers the historical problem as well. It seems reasonable to assume, following Bellman, [2] that optimum scheduling can be determined for any starting date, so that our problem is reduced to selecting between alternative optimally scheduled starting dates, where each has an identifiable stream of social benefits and costs.

CRITERIA FOR OPTIMAL CHOICE

Our basic objective is the maximization of (domestic) social welfare, defined as the sum of the present values of the difference between social benefits and social costs during the operational time period of the project. Specifically, it may be defined as the sum of producers' surpluses and consumers' surpluses which accrue to the relevant members of the community,[3] plus or minus any external economies created for or imposed upon members of the relevant group who are neither producers nor consumers. We have discussed this criterion at greater length [4] and believe it to be in general conformity with established tenets of welfare economics. [5] The optimal schedule for any project is simply that which, taking externalities into account, maximizes the excess of benefits over costs for the given starting date. Where production in one period can be increased only at the expense of production in another period (because of resource exhaustibility),this criterion means production should be postponed only to the extent that savings in costs (resulting from a lower output rate), or increases in value (resulting from higher demand in subsequent periods) produce in the subsequent period net benefits whose present value exceeds the net benefit of expanded current production. [6]

The historical timing problem in this context produces an equally simple decision rule, as follows: all projects should be undertaken immediately they show a positive net present value of benefits minus costs, unless it can be shown that the net present value will be increased; delay should be imposed where current net present values can be increased to the point where additional delay ceases to provide any increase.

Most of the capital budgeting and benefit-cost literature suggests that welfare is maximized when all projects with positive net benefits are incorporated in the current capital programme.[7] The need for explicit consideration of the opportunity afforded by possible delay appears to have been first recognized by Gordon [8],although the appropriate delay criterion was advanced by one of us in another context. [9] It is merely a question of comparing the present value now of proceeding immediately with that of proceeding one planning period ahead. If the latter is larger, the project should be delayed; if it is not, it should be proceeded with immediately (assuming it is worth doing in the first place). The decision to postpone can be made on a one-period-at-a-time basis; there is no need to try to specify precisely how long development should be delayed. The resulting set of projects, after projects which benefit from postponement have been removed, is the set whose development will maximize social welfare.

The cyclical timing problem arises because it is not possible to undertake the entire welfare maximizing set within the planning period. The existence of constraints makes it impossible to attain the welfare maximum. While attention should be directed toward devices to ease the constraints, as long as they remain binding, it will be necessary to postpone some of the projects.

Solutions to the problem of selecting projects for postponement have been examined within a mathematical programming framework by Weingartner,[10] while one of the authors has developed an index of postponability which can be used to select for postponement those projects which enable the constraint to be met with a minimum sacrifice in welfare. [11] Basically this index has as numerator the reduction in net present value of benefits resulting from postponement and as denominator, a measure of the activity or input which is constrained. Projects with low values are successively eliminated until the constraint is satisfied.

DECISION STRUCTURE AND THE HISTORICAL TIMING DECISION

While the basic decision making structure in our economy remains a decentralized one in which decisions are taken by competing units in response to market forces, it would be inaccurate to describe it as one in which government played no role. In many instances, government has a primary ownership interest in the resource; in others, the exercise of police or taxing power of the state may have an important influence on the decisions made by market participants. The basic question we must ask is whether the decisions that result in this system provide a tolerable approximation to ideal welfare maximizing decisions, or whether they would be improved by altering government practice. No one doubts that they could be improved, in an abstract sense, by an omniscient and omnipotent agency; we must live, however, in a world populated by real corporate managements, real governments, and real bureaucracies, which are neither omniscient nor omnipotent.

There is impressive evidence, both theoretical and empirical, that decentralized decision making systems are generally efficient, since they tend to minimize the need for data collection and manipulation by the participants. It is at least sufficiently impressive that certain socialist societies, in which private ownership has been abolished, have recreated the market mechanism as a means of improving the planning machinery. [12] Despite its impressive credentials, there has been broadly based concern over the ability of the market mechanism to produce fully satisfactory decisions where resource development is involved. This concern has arisen primarily on two grounds. Private developers frequently do not, and often cannot (because of competitive pressures), take externalities into account in the planning process. As a consequence, projects generating external diseconomies are too likely, and those generating external economies are too unlikely to be accepted and undertaken. Decisions made on the present value basis are sensitive to the selection of interest rates at which discounting is performed. Private sector capital costs are often, it is alleged, too high, with the consequence that short lived alternatives are chosen over longer lived alternatives, and an anticonservation bias is imposed on the decision making process.

We do not propose to examine the question of externalities here, except to comment that public agencies have, on the whole, a track record at least as deficient as that of the private sector in this regard. While there is a presumption that some kind of collective action will be needed to deal with the problem, it may require as extreme an alteration in public sector thinking and behaviour as anywhere else. We leave this area, however, on the ground that it is not really part of the timing problem; in any event it has been dealt with adequately by others. [13]

The essence of the timing problem, or at least of the historical timing problem, is that of ensuring that the discount rate used for the evaluations of individual projects is in some sense appropriate. Will private decisions be based on evaluations made at discount rates which are excessive in relation to the appropriate social rate of discount, and if they are, what of it?

Unfortunately, we have no definitive answer. The issue revolves around the question of what constitutes an appropriate social rate of discount, and this is not one of the questions on which economists have developed a significant measure of agreement. In our view this is not too surprising, since the problem is a philosophical one, with its roots set firmly in the age-old controversy over universals. Is *society* merely a collective noun applied solely to the existing members of the community, or is it something greater, embracing a Burkean continuum involving not only the present generation but also past generations and generations yet unborn? While some economists have been philosophers, notably Smith and Mill, most are not and are poorly equipped to handle the question. In any event, the philosophers do not seem to have gone much beyond taking sides, so we cannot expect much guidance from that direction. If we accept the nominalist position, there appears to be little difficulty in following Arrow [14] in arguing that benefits flow only to individuals, not to an abstraction called society, and that the relevant discount rate for evaluating the benefits flowing to individuals is the rate at which they make their own consumption and investment decisions. If so, the private sector discount rate is basically appropriate.

If on the other hand, we accept an organic theory of the state, we may be tempted into following Marglin and argue that, in situations where benefits to future generations are involved, such as those with which we are dealing, people will prefer a lower rate than that usually used for their own decisions, providing it is imposed on all segments of society making such decisions as well as on themselves.[15] A similar argument has been advanced by Baumol.[16] This is, of course, not the only position compatible with the organic theory of the state; for example, the accepted Marxist position seems to be that individual preferences are totally irrelevant and that society's investment decisions ought to be taken purely on the basis of achieving efficient use of investment funds, as argued by Dobb.[17] This implies the equalization of the marginal efficiency of capital in all uses, and, applied to a decentralized decision making system, it would mean the use of a discount rate which is more or less the equivalent of the private sector rate in our society. Of course, many adherents of the Marglin position would argue that existing Marxist governed societies have done at least as bad a job of conserving resources for the future as have their nominally capitalistic counterparts, a view which we are prepared to accept.

Maybe, however, the irreconcilable philosophical conflict is as irrelevant as many critics of philosophy have claimed. What are the practical consequences? If Marglin's analysis is correct, the major defect of private sector decisions is that they are based on an excessive discount rate. This leads to an underestimation in the decision making process of future benefits relative to present costs and of future costs relative to present benefits. This has two results. First of all, it will lead to projects which are accepted being somewhat present oriented, with perhaps insufficient provision for the future in terms of benefit scheduling. But it also results in the rejection or postponement of projects whose benefits extend into the future, but whose current benefits are foregone because the project will not produce an adequate rate of return at the present time. In this respect, it is perhaps biased toward making excessive provision for the future. Whether the impact of the possible antifuture bias in projects undertaken is greater or less than that of the profuture bias in projects rejected is obviously an empirical question on which we have little evidence; data on developments not undertaken are remarkably hard to come by. The only evidence we can offer is the record of water resource development in the United States, where the agencies largely responsible, the Corps of Engineers and the Bureau of Reclamation, were largely freed from the need to reject projects not meeting private sector profitability requirements. Many of these projects appear to be justified by benefits which extend beyond the future into the hereafter. [18] Whether the resulting conservation orientation is precisely what conservation minded groups had in mind is not exactly clear. We suggest that the evidence in this area supports the view that raising the required rate of return does more for the future by discouraging presently marginal projects than it harms it by leading to inadequate provision for the future in projects which are forced to meet the higher private sector discount rate. The "if it flows, dam it" philosophy is a product of artificially low, rather than excessively high, discount rates. When government undertakes an investment that the private sector is "incapable" of undertaking or unwilling to undertake, it is only too often an investment that nobody in his right mind should have undertaken.

What has been forgotten is that the "faulty telescopic faculty," of which Pigou complained so long ago [19] is inherent in all of us and not confined to private sector executives. The appropriate comparison group for the latter is not some abstract philosopher-king but the real live cabinet minister. The forces which produce short-sightedness in the former are also operative on the latter. Casual observation suggests that the job tenure of the former is significantly more secure than that of the cabinet minister, and that he is freer to take the long view than is the latter, whose relevant time horizon may not extend beyond the next election or even the next vote of confidence. Even statesmen, certified by the subsequent verdict of history for surpassing these limits, tend to be remembered for isolated, rather than consistent, behaviour in this regard.

Given that the general private sector interest rate may be appropriate for resource development decisions, or that if it errs, it is because it is too high,we might inquire as to whether the institutional framework surrounding the resource industries is such that this is in fact the effective rate on which historical timing decisions are based. This involves questions of taxation, tenure, and regulation.

In discussing taxation, attention tends to be focused on the special provisions with respect to certain resource industries found in the Income Tax Act, on the dramatically higher provincial levies imposed on certain of the resource industries, and on the federal-provincial squabble over revenues shares, which has raised marginal tax rates to levels beyond 100 per cent in certain circumstances, a situation likened by one observer to "two hogs fighting under a blanket for the same acorn." It is perhaps more appropriate to consider provincial taxation as part of the overall tenure question and to confine our brief remarks to the corporate income tax.

A further caveat is that our concern is with corporate tax legislation as applied in Canada at the present time. Canadian tax legislation has always differed significantly from that of the United States, even if certain items have had similar names. An American debate about the rights and wrongs of their depletion allowance dumped into the Canadian intellectual marketplace as a free good has unfortunately led some of our more advanced thinkers to ignore our own system before pronouncing judgment thereon. If there were a generalized subsidy to resource development that did more than offset certain peculiar disabilities which we discuss below, there would be a tendency for development decisions to be made at an effective rate of return rather less than that typical in the private sector. This is not necessarily bad, if one accepts Marglin's position that the relevant discount rate should be lowered.

The principal "special" provisions for the mineral industry in Canadian tax legislation, for example (ignoring certain items scheduled to disappear), include the right to expense certain costs incurred in the acquisition of legal rights to develop resources, provisions with respect to the deduction of development costs and the so-called depletion allowance.

Rights to develop resources have, unlike many interests in land, a finite duration, and charging them to expense during the life of the rights so acquired would appear to be a perfectly normal accounting procedure. However, there is abundant evidence that the tax status of such payments is capitalized in the process of bidding for such rights, and that the provision enhances provincial revenues rather than doing anything for the purchaser of such rights.

Present tax legislation also permits the deduction of intangible development costs at rates which appear, in many instances, to be rather more rapid than the decline in useful value of the assets to which they are related. For example, costs of drilling oil wells may be deducted as incurred, even if the resulting well is productive. There are a variety of such intangibles incurred in other industries(for example, interest during construction, R&D expenditures, patent licence fees, payments in respect of goodwill, costs of engineering studies for projects that never come to fruition, and so on).

There are many inconsistencies in the tax treatment of such items, but the right to expense the type of expenditure is by no means confined to the resource industries. Where the line should be drawn is a practical matter; unless and until there are unlimited carry-forward and carry-back provisions in tax legislation, there is much to be said for a provision that enables these expenditures to be claimed at a time when there is enough income available.

The result may well be some tendency to reduce both income and assets in comparison with those which would result under a hypothetical and perfect accounting system. Similar results no doubt occur in many other industries because of discrepancies between capital cost allowances and economic depreciation. In their defence, it must be noted that the corporate income tax is not neutral among industries which differ in capital intensity and/or risk, and that it systematically discriminates against industries which are more capital intensive or more risky than average To the extent that so-called excessive capital cost allowances (including those at 100 per cent rates) serve to offset this bias, they serve to improve economic efficiency rather than detract from it.

Last, but not least, in our discussion of unusual tax items is the depletion allowance. The availability of depletion allowances in Canada has always been restricted, and they have never been available as a "laundry" facility for the conversion of high-bracket personal income into tax-free dollars as they are in the United States. We appear to have met our "laundry" requirements in other ways. Among the more notable of these is the channelling of funds through the production of blue movies, reflecting our collective passion for the arts. Our depletion allowance is a tame affair, essentially permitting 133 per cent of exploration costs as incurred to be written off. It is quite clearly a subsidy to exploration. If the fruits of exploration accrued solely to the explorer, such a subsidy would be hard to defend. But they do not. Mineral deposits themselves frequently extend beyond the boundaries of the exploring companies' property and under that held by other companies or the Crown. And information produced by unsuccessful exploration in one area may hold the key which leads to success somewhere else. In short, there are substantial external benefits generated by exploration, and the case for a subsidy on exploration is analytically identical to that for a tax on pollution.

To summarize our excessively brief but unduly protracted discussion of taxation, there appear to be valid reasons behind some of the tax provisions peculiar to the resource industries. Whether the specific provisions are too generous or the reverse requires detailed empirical study rather than the rhetoric which has been the chief contribution of both sides in the debate for many years.

Before moving on to discuss tenure related problems, it is perhaps pertinent to note an additional dimension of the tax environment, centred not on the actual provisions of tax legislation but on the predictability of future tax liability. Most resource development investments have comparatively long lives, and conservationists are concerned that they are made on a basis which makes adequate provision for the longer-term future. Investment decision making must be based on an evaluation of benefits and costs over the life of the project. Forecasting such benefits is intrinsically difficult in any event, but it is rendered doubly so if uncertainty about future tax rates increases the dispersion of probability distributions of future net benefits. The entire economy has gone through a half decade or more of massive uncertainty as a consequence of tax reform. Even in those areas where the process has been completed, substantial uncertainty will remain until a body of case law on the new Income Tax Act has been developed. In the case of many of the resource industries, the process of reform has been even more protracted, and major revisions were still pending in September 1974. Prior to tax reform, the resource industries enjoyed a fairly stable tax climate for twenty years or more. The entire exercise has injected an element of political risk into the decision making process, which has had an inevitable price in terms of the risk premiums required on resource investments. While this has probably reduced the rate of investment in resource development and aided conservation, it has imposed the bias in favour of short lived rapidly depleting projects that one might expect from a rate of discount which is excessive in social terms.

Tenure has also an important influence on the decisions that affect the timing of resource development. The reasons are essentially the same. The terms on which the right to exploit resources are granted affect the net benefits accruing to the operator. Adam Smith's observation that short-term tenants are unlikely to make improvements on the land and are even likely to let existing improvements fall into disrepair remains as valid today as when it was written two hundred years ago. It is simply unrealistic to expect the developer of a resource property to do anything to enhance the value of the property after his tenure has expired. Where tenure can be made indefinitely long, as long as production continues (as it is in the typical oil and gas lease), this basic problem is avoided, though there are related problems which arise in attempting to ensure the restoration of strip mining sites, for example. Fixed-term concessions impose a short-term bias, even if the initial term is long. A variety of problems have arisen, for example, in Venezuela as the expiry date of oil concessions approaches.

What is perhaps more significant than the continuity of formal tenure however, is the question of terms. What has become increasingly clear is that resource developers have become in fact, if not in law, tenants-at-will of resource owning governments. Lease or concession terms have been unilaterally amended by host governments, with no regard for previously agreed renegotiation timetables. Such timetables are bad enough, since they increase uncertainty about the value of what is left after the renegotiation date. In many provinces today, there is real doubt about the value of what will be left after Christmas, and this cannot but affect investment planning in an adverse manner.

We are quite aware of the proffered justification for changes in oil and gas rental and royalty rates—that they were imposed to capture windfall profits arising out of international price changes. Even if one accepts that it is government's right and duty to seize all windfall gains, with or without any responsibility for making up windfall losses, the explanation is still simply unacceptable, because many of the increases in royalty rates were initiated prior to the events that led to the major price increases. There is, of course, the possibility that the increases were foreseen by government. But there is an equal possibility that they were foreseen by industry as well. Investment commitments may have been entered into on the basis of expected higher prices on which profits would be merely normal and involve no windfall element. Expenditures on tar sands development and on offshore exploration in the Arctic would both appear to fall into this category, since neither could have appeared profitable under the price prevailing two years ago. Instead of reducing a windfall profit to normal levels, changes in royalty rates may have turned a normal profit into a windfall loss. The effect cannot be other than to discourage long-term planning to anticipate long-term requirements. Again the effect is to make the market solutions less rather than more responsive to social needs.

Timing decisions can be affected not only by government's control over the terms on which tenure is granted, but also by its control over the rate at which potential resources are made available to would-be developers. In some cases, we have had a policy of making resources available on demand, permitting development by anyone who files a claim or applies for a permit. In other cases, whole areas have been withheld from availability for indefinite periods, as the United States has held Naval Petroleum Reserves out of the development stream.[20] The major Canadian examples have been the provincial reserves set up by Alberta soon after it acquired title to its natural resources. Most of these were eventually developed, although the British Block, which has substantial gas reserves, has been held back not so much by provincial action as by the accident of occupation by the defence research establishment at Suffield. There have been proposals that similar reserves be set up elsewhere. Establishing reserves of this type leads to substituting administrative judgment for market forces in determining the pace of development.

Policies of open availability have typically, in North American experience, been accompanied by a requirement that the acquiring company
proceed with exploration and development forthwith. One object is to
prevent monopolization, which can be taken care of in a variety of other
ways; another is to discourage speculative acquisition for future use. It is
fairly clear that the open availability system, with a forced development
pace, will encourage development at the earliest moment it becomes
economically feasible and prevent the holding of the resource off the
market in the expectation of better prices. In short, the system contains an
inbuilt bias toward premature development.

Because of this, there is an attraction in the system of withholding
reserves. In principle, these can be held off the market until their value, and
the prospective contribution to social welfare, is maximized. More
precisely, development should be postponed until the rate of increase in the
present value of net benefits drops below the social rate of discount. There
is, unfortunately, not much evidence that this is how administrators will
place resources in the hands of developers. They will either hold them off
the market forever for fear of being accused of a giveaway, as has
apparently been the case with the Naval Petroleum Reserves in the United
States, or dispose of them as soon as the prospect of an attractive addition
to revenues appears, because of the short-term bias we have attributed to
the political decision maker. The British Block is the only provincial reserve
of significance remaining in Alberta, and it has remained on the shelf for
reasons unrelated to resource development as we have noted. Most of the
others were fed into the development process at a time when the oil market
was glutted, when their output could be sold only by cutting back
allowables on other production. (On the other hand, at least some of the
prices received might well have been greater than anyone would have paid
subsequently, even today.) What we are really asking the administrators to
do in this situation is to function as efficient speculators; and this is a role
for which civil servants are usually unsuited whether by temperament,
training, or tradition.

Efficient speculation would improve the historical timing of resource
development. One way of providing such efficient speculation would be, it
seems, to permit private acquisition on demand, after public auction, but to
remove those provisions requiring expeditious development. There might be
problems in developing a tax system which secured a satisfactory share of
speculative profits for the public treasury, but we note that few such holders
would qualify for capital gains treatment, and normal tax rates would
apply. The problem may be less serious than it seems.

Direct regulation of resource development projects has also been extensively employed with respect to water resource development, tar sands and

pipeline projects, among others. These have been effective in postponing certain projects. It is by no means clear, however, that this has been an effective device for securing optimum postponement. Most of the agencies operate under statutes that require them to consider economic feasibility as well as physical or environmental criteria under a blanket requirement that they consider the public interest. In general, the economic criteria that have been applied have not been concerned with optimum timing, merely with attempting to ensure that the project will be economically feasible. Typically, this means that projects can be approved as soon as they meet the feasibility criterion without any concern over whether it might be better to wait. As practised, the whole process is once again biased toward premature development.

   In summary, looking at the question of historical timing, there is very little evidence to suggest that intervention by public authorities will improve the timing of resource development decisions over those which would result if market forces were allowed to operate without the premature development bias implicit in present tenure regulations.

THE CYCLICAL TIMING DECISION

   The cyclical timing problem may be another story. Historically, the unfettered market has demonstrated a tendency to "bunch up" investment projects, including those involving natural resources. This bunching up has been a major factor in creating business cycles. There is concern in at least some circles over the possibility of a boom-bust sequence initiated by a possibly excessive bunching up of resource projects in the latter part of this decade. Since many of the projects are in areas subject to regulation, it is almost certain that we will not rely on the free play of market forces to determine which project ought to be postponed.

   The market would resolve the problem in its own way by letting interest rates and the prices of critical inputs rise until enough projects disappeared to permit us to digest the rest. The chief problems with this approach are that the relevant prices (many are specific wage rates) are not flexible in both directions, that a burst of inflation might be required to rearrange appropriate relative prices after the bottleneck had passed, and that cyclically high interest rates would impose a bias in favour of shorter lived projects. There is, at least *prima facie*, a case for the exercise of some administrative discretion as a substitute for a process of this type.

   What guidelines should be employed? To what extent should the resource sector be permitted to grow at the expense of other sectors of the economy by attracting resources away from them? We do not think there should be any arbitrary limit. If potential investment pressures are too great, non-resource projects are as good candidates for postponement or abandonment as resource projects. This implies that there may need to be direct controls on all capital projects over a certain size. An attempt should be made to identify critical supply areas, to determine what the effective availability constraints are and the extent to which they can be eased. All projects should then be analyzed in terms of their requirements of the scarce inputs. A selection among the projects can be made by considering the postponement cost associated with individual projects. This problem lends itself to a mathematical programming solution in which postponement costs are minimized subject to the constraint on scarce inputs. This problem, in a somewhat narrower setting, has already been analyzed.[21]

Notes

1. R.L. Gordon, "A Reinterpretation of the Pure Theory of Exhaustion," *Journal of Political Economy* 75 (1967): 224-86; H. Hotelling, "The Economics of Exhaustible Resources," *Journal of Political Economy* 39 (1931): 137-75; G.D. Quirin and B.A. Kalymon, "Conservation and the Optimum Exploitation of Exhaustible Resources," Working Paper No. 73-03, Faculty of Management Studies, University of Toronto, 1973.

2. R. Bellman, *Dynamic Programming* (Princeton: Princeton University Press, 1957).

3. This may be defined on the basis of residence or citizenship, a choice which is obviously dependant on concepts of government responsibility. Recent legislation, e.g., the Foreign Investment Review Act, has tended to treat landed immigrants as within, and non-resident citizens as without, the community to which government is responsible.

4. G.D. Quirin, "Energy: Policy Choices Regarding Exploitation and Export," in *Proceedings of the Royal Society of Canada Symposium on Energy* (Ottawa, 1973); and G.D. Quirin, *The Capital Expenditure Decision* (Homewood, Ill.: Richard D. Irwin, 1967).

5. W.J. Baumol, *Welfare Economics and the Theory of the State*, 2nd edition (Cambridge, Mass.: Harvard University Press, 1965); and M. Dobb, *Welfare Economics and the Economics of Socialism* (Cambridge: Cambridge University Press, 1969).

6. A.D. Scott, *Natural Resources: The Economics of Conservation* (Toronto: University of Toronto Press, 1955).

7. S.A. Lutz and V. Lutz, *Theory of Investment of the Firm* (Princeton: Princeton University Press, 1951).

8. M.J. Gordon, "The Optimal Timing of Capital Expenditures" Working Paper, Faculty of Management Studies, University of Toronto, 1973.

146   *Anthony D. Scott*

9. G.D. Quirin, "Energy," pp. 181-85.
10. H.M. Weingartner, *Mathematical Programming and the Analysis of Capital Budgeting Problems* (Englewood Cliffs, N.J.: Prentice-Hall, 1963).
11. G.D. Quirin, "Energy."
12. O. Sik, "Socialist Market Relations and Planning," in *Socialism, Capitalism and Economic Growth*, ed. C.H. Feinstein (Cambridge: Cambridge University Press, 1967).
13. J.H. Dales, *Pollution, Property and Prices*(Toronto: University of Toronto Press, 1968).
14. K.J. Arrow and R.C. Lind, "Uncertainty and the Evaluation of Public Investment Decisions," *American Economic Review* (1970): 364-78.
15. A. Maass, et al., *Design of Water Resource Systems* (Cambridge, Mass: Harvard University Press, 1962).
16. Baumol, *Welfare Economics.*
17. Ibid.
18. J. Hirshleifer, J.C. DeHaven, and J.W. Milliman, *Water Supply, Economics, Technology and Policy* (Chicago: Univeristy of Chicago Press, 1960); J.V. Krutilla and O. Eckstein, *Multiple Purpose River Development* (Baltimore: Johns Hopkins University Press, 1958); and A. Maass, ibid.
19. A.C. Pigou, *Economics of Welfare*, 4th edition (London: Macmillan, 1932).
20. Not entirely. Some portions of the Elk Hills reserve have been developed to protect against drainage from contiguous private leases.
21. H.M. Weingartner, *Mathematical Programming,* and G.D. Quirin, "Energy."

Comment

ANTHONY D. SCOTT

This article, I must confess, irritated me more than it informed me, because it deals with what I believe is the wrong subject for this meeting. To a greater or lesser extent, the authors are concerned with the allocation of minerals over time. They assure us that the market place has this problem in hand and at the same time alert us to dangers that imperfections and failures in the market mechanisms, stemming from a range of sources (mostly governmental), might result in a time path that was not the best. Orthodox analysis often conveys fears that other private market flaws and failures may land future societies in difficulties; but Quirin and Kalymon are not too worried.

Now this important question (about the path of consumption over time) is not the one that faces British Columbia as a metals producer or Canada as an energy producer. The implied advice that output be such that price equals incremental operating plus user costs, will contribute to an objective that this province does not knowingly seek. What this province does want to know is the optimal dating of exploration, mining, and selling ores, bearing in mind that such events imply the rise and decline of mining towns, railways, and ports and, more indirectly, the rise and decline of total West Coast population and per capita income.

But the challenge to the economist is twofold. First: what objectives should a province seek? Presumably the answer lies in some choice between the time paths of total or per capita incomes. With migration as expensive as it is today, it is never clear which choice politicians should worry about. In any case, income would include both factor earnings and provincial tax and royalty revenues. Second: does this time path coincide with that which British Columbia would follow if it accepted the Quirin-Kalymon advice and acted as a landlord maximizing the net present value of the flow of future minerals? That is, if one is concerned about urbanization, migration, and the social structure, is maximization of the present value of rent the best plan?

Perhaps it is. A traditional strategy of economic advisors faced with difficult distributional questions has been to counsel maximization of national income, so that any who suffered could be recompensed by adjustment and redistributional policies until everyone gained. Our problem concerns both distribution and allocation over time. By analogy, therefore, we might advise British Columbia to make hay while the sun shines, putting much of the hay in the barn to feed those who are attracted to British Columbia by the stocks of hay, those who are hurt by the rapidity of the haymaking, and the future descendants of the first two groups, wherever they choose to live. This strategy would at least be better than trying to grow hay in the wrong seasons or having good hay left in the barn after all the haymakers have died or left. But there may be even better strategies for reconciling possible population, urban, emigration, and infrastructure growth paths with alternative exploration, development, and secondary recovery and closure sequences for separate mines and whole mineral regions. These may be achieved as a byproduct of searching for the global optimum for consumers as a whole. But it seems to me more plausible that we will require more *ad hoc* or what Quirin and Kalymon call "cyclical" policies to space development and population over time and over the landscape.

In my *Natural Resources* book [1], I referred to all this as the "ghost-town" problem. It comes up repeatedly, whenever new mineral regions realize that they have control over "the problem of timing in resource development," to use Quirin and Kalymon's title. Norway and the Canadian Northwest Territories are both thinking it out today. Can we be sure that they need merely follow our authors' advice to avoid unduly distorting changes in tenure and taxation? The problem is more difficult, I think. I recommend the simulation sequences developed by John Helliwell and others for arctic gas[2] for a more penetrating view of the timing problems and opportunities open to exporting governments.

Notes

[1] A.D. Scott, *Natural Resources: The Economics of Conservation* (Toronto: McClelland & Stewart, 1973).

[2] P.H. Pearse, ed., *The Mackenzie Pipeline: Arctic Gas and Canadian Energy Policy* (Toronto: McClelland & Stewart, 1974).

The Government Role
in Mineral Exploration

FREDERICK M. PETERSON

Mineral exploration deserves special attention today as a way to relieve the shortages of oil, gas, and hardrock minerals. In this article, it is argued that the government should perform or contract out the early exploratory work on public lands because information spillovers, scale economies, and risk aversion prevent the market from functioning properly. Failure of the market encourages the domination of large, international firms, reduces the rate of exploration, and subtracts from government mineral revenues. If the government is not prepared to enter the exploration business, the same arguments can be used to support less direct government involvement, such as regulating the industry and sharing in profits and risks. The discussion will focus on British Columbia, but the conclusions apply to any region whose mineral resources cover large areas and are relatively unexplored.

Government involvement in mineral exploration is not a new concept. The British Columbia Department of Mines and Petroleum Resources provides analytical and assay services, performs field work, publishes aeromagnetic surveys, builds and repairs roads and trails into mining areas, and disburses grubstakes to qualified prospectors to encourage exploration.[1] Senator Jackson of the United States has proposed federal exploration and development of the Outer Continental Shelf and Naval Petroleum Reserves,[2] but that would not be anything new. The navy has already explored many of its petroleum reserves, and the federal government has subsidized or peformed mineral exploration through several programmes in recent years, starting with the Strategic Materials Development Program, which spent $30 million between 1939 and 1949 looking for certain scarce metals, and ending with the Office of Mineral Exploration, which has operated since 1958. Preston[3] feels that these programmes have stimulated small-scale operations and have not produced many reserves. He sees the need for large-scale operations using modern exploratory techniques.

ARGUMENTS FOR GOVERNMENT INVOLVEMENT: MARKET FAILURE

The case is strong for government involvement in mineral exploration, at least on public lands. The arguments resemble those used, quite correctly, to justify government involvement in research and development. Like research and development, mineral exploration generates widespread benefits, involves returns to scale, and requires risk taking, which prevents markets from operating efficiently. Neither research and development nor mineral exploration is conducted at optimal levels without government intervention. As the research becomes more basic and the exploration more pioneering, the market becomes more inhibited. Just as the government performs and subsidizes research and development, it should perform the early exploration on its own mineral properties, or intervene in some other way. By so doing, it would correct the problems of information spillovers, scale economies, and risk aversion, and generate benefits in the form of greater competition, higher rates of exploration, and increased mineral revenues.

Information Spillovers

The benefits from mineral exploration spread widely to payers and nonpayers alike. Mineral deposits on one piece of land are related to deposits on another, so that exploration activities yield information that extends well beyond the piece of land being explored. In a true market situation, the mineral explorer seldom owns the mineral rights on all the affected land. Unless he can make deals with surrounding mineral owners before proceeding or withhold the information generated for later use or sale, he will lose some of the benefits of his activities and delay or cancel some worthy exploration ventures.

There are many examples of the information spillover effect, especially in oil and gas exploration. Colonel Drake completed the first oil well in Western Pennsylvania and provided information that started an oil boom. Many people made millions of dollars, but he failed to make enough to pay royalties and died in poverty.[4] Miller presents similar evidence from the Rocky Mountain and Southwestern areas of the United States.[5] McKie talks about trend leasing, the strategy by which free riders race to land offices and capitalize on exploratory information generated by others.[6] In another paper, I discuss the skyrocketing lease prices that followed Atlantic Richfield's giant oil discovery on the North Slope of Alaska.[7] Dale Jordan's article contained in this volume describes the difficulty of getting petroleum companies to explore areas that have been broken up by British Columbia's chequerboard leasing programme. At least part of the problem is that they are reluctant to provide free information to the parties who already hold tracts. In the area of gold exploration, John Sutter touched off the California gold rush by digging a mill race, but he got little of the gold. In British Columbia, the discovery of gold at the mouth of the Fraser River in 1858 led men all the way up the river to the fabulous deposits of the Cariboo.[8] The benefits of mineral discoveries are widespread indeed.

To capture some of the spillover benefits, prospecting and exploring companies often make deals with surrounding leaseholders or engage in cooperative ventures with them. In the oil business, companies sometimes exchange drilling information, and, at other times, they arrange Coase-type side payments called "bottom hole money" and "dry hole money" to internalize the information spillovers.[9] Cooperative ventures involve the sharing of the costs and/or proceeds from an exploratory operation. The actual exploring is often done by a small, independent company, which affiliates with a major oil company that holds many leases. Side payments and cooperative ventures are hard to arrange when numerous lease holders are involved. Miller argues that fragmented ownership of the surrounding leases raises transaction costs to the point where useful projects have to be scrapped.[10] The companies also might go their separate ways and wastefully duplicate exploratory activities.

Many companies make drastic but futile efforts to conceal information about their exploratory activities. Petroleum companies have roaming oil scouts observing other companies' drilling operations. They reportedly can determine the depth of a well and the type of geological formation by counting the lengths of pipe and observing the colour of the discharged drilling mud. They even use auto licence lists to keep track of their competitors, and they maintain informers in county land offices to learn about leasing activity.[11] With all these efforts and the basic difficulty of concealing and selling information, an individual exploration company cannot capture all the benefits derived from mineral search, which means that the market generates less exploratory activity than is optimal.

Some contributors to this volume have argued against generalizing the spillover effect, because some types of deposits are not spatially related to one another, and because, they claim, spillovers apply more to sedimentary, alluvial, and stratabound deposits than to hardrock minerals like nickel and copper, whose deposits are randomly and widely distributed. In effect they were saying that finding nickel and copper at one site bears little indication of nickel and copper being found elsewhere in the vicinity. If that is so, how do they explain the rush to stake claims around Texas Gulf Sulfur's nickel-copper discoveries in Ontario? The stakers, at least, think that such deposits are spatially correlated. I concede that information spillovers vary in importance from mineral to mineral but feel that in no case are they unimportant.

Economies of Scale

Scale economies in both the production and dissemination of exploratory information further distort the market. Large-scale exploratory techniques favour big firms and, when exploratory information is sold, marginal cost pricing is discouraged because profits would be negative.

The production of exploratory information often requires expensive equipment and sophisticated geophysical techniques. Offshore oil rigs are enormous, but they cannot be built smaller if they are to operate in a given depth of water. Geophysical techniques have high fixed costs but low incremental costs for exploring an extra square mile. It appears that such scale economies are gaining in importance and are squeezing out small operators.

Take the case of hardrock minerals. Grubstakers and other prospectors have played an important role in British Columbia,[12] but their role in finding these minerals may decline as high grade surface outcrops are depleted and lowgrade or hidden deposits remain. These require airborne magnetometers, gravimeters, and cameras that have scale economies.[13] Once a plane is airborne, it costs very little to fly it over another tract of land. Big companies can better afford to explore large areas and employ the people needed to interpret the results. The United States appears to be well down this road as the number of mineral patents issued dropped from 16,000 per decade at the turn of the century to less than 1,000 by 1940.[14] In the words of the United States National Science Foundation:

> Technologically, [the prospector's range] has been cut down because he cannot afford the equipment with which modern exploration must be carried on. Practically, he has worked himself out of a job in the metallic field because he has discovered virtually every worthwhile deposit that can be found by visual and simple instrumental methods within the confines of the United States.[15]

The United States has gone even further with oil and gas, which small firms once found in abundance by looking, guessing, and drilling—without sophisticated geophysical data. Colonel Drake chose his site near oil seepages and drilled only sixty-nine feet to make his famous discovery in 1859.[16] Slichter claims that surface geological methods of searching for oil were nearly finished in the United States by 1924. By 1930, in the Gulf Coast of Texas and Louisiana, virtually every salt dome had been discovered that was shallow enough to be discovered by seismic refraction.[17] As time goes on, larger and more sophisticated efforts will be required to find oil, gas, and hardrock mineral deposits in British Columbia. According to Shearer, small Canadian firms already have trouble competing in oil and gas.[18] Similar problems with hardrock minerals might be coming.

Independent service companies perform much geophysical work under contract, but they do not help the small producing companies as much as one would expect. For one reason or another, big companies make the greatest use of their services. They may get better deals because of scale economies and purchasing power, or the small companies may lack the sophistication needed to interpret the information once it is purchased. It is also possible that the small companies can buy and use the services on equal standing, but they go after a kind of deposit that does not require such geophysical measurements.

In any case, scale economies in the dissemination of exploratory information prevent exploration service companies from pricing efficiently. After paying the initial fixed cost of acquiring exploratory information, these companies incur negligible costs in running off another copy and selling it to someone. Market efficiency would require them to set the price of their information near zero, but they would go broke with such a low price. Exploration services is a classic decreasing cost industry, like Musgrave's example of the bridge whose optimal tolls are zero.[19] Just as governments build and operate bridges, they might generate and disseminate exploratory information for a nominal fee in order to utilize scale economies and distribute the information efficiently.

Risk Aversion

Risk aversion is often thought to be a problem with mineral leasing; and Leland, Norgaard, and Pearson have devoted a paper to the problem.[20] They assume risk aversion and proceed with the remedies, but, given the amount of disagreement which exists surrounding this topic, I will try to explain risk aversion and develop some evidence of its presence, rather than just assume it.

If businessmen are risk averse and mineral exploration is one of the more risky businesses, then the market would generate less exploration than is optimal and government involvement would be justified. A risk averse person is one who rejects an even gamble; it hurts him more to lose than it pleases him to win. This puzzling aspect of human behaviour was explained in terms of expected utility maximization in 1731 by Daniel Bernoulli.[21] His theory says that a decision maker gets less and less utility out of each additional increment of income and that, when faced with risky choices, he acts to maximize his expected utility rather than his expected income. To such a decision maker, an exploration project with positive expected income to him and to society could have negative expected utility. Such a project would not be undertaken in a free market system. That is why government intervention might be justified.

There are debates about whether firms are risk averse and whether small firms are more risk averse than large ones. These are important empirical questions that have not been decisively answered, but the following can be said. For a firm to be risk averse, its utility function need only exhibit diminishing marginal utility in the relevant range, a condition that has been established by at least one empirical study. Grayson conducted a von Neumann-Morgenstern experiment with nine independent oil operators and found that their utility functions had diminishing marginal utility.[22] He did not compare big firms to small firms; so the question of who is more risk averse was not answered directly. However, most of Grayson's subjects showed a strong reluctance to lose more than $200,000, probably because it would bankrupt them. It is safe to say that Exxon, with its hundreds of millions spent on exploration, would not shy at a $200,000 gamble, even in Grayson's game, with its hypothetical questions. A respectable exploration programme, of course, costs considerably more than that. One exploration venture might cost a million dollars and have a 10 per cent chance of success. After financing ten such ventures, a company would still have a 35 per cent chance of finding nothing. Only a large company would take such a gamble.

Risk aversion itself would not justify government involvement in mineral exploration. If all industries were equally risk averse and faced about the same level of risk, the government would not necessarily improve things by correcting the risk aversion in mineral exploration while leaving risk in other industries alone.[23] It is only hypothesis, but I think that mineral exploration is one of the more risky businesses. In oil, for instance, exploration ventures are expensive, and success ratios are low. The cost of drilling varies widely according to the depth achieved and difficulties encountered, and the payoffs vary from a billion dollars to a worthless hole in the ground. Recently, the risk problem has been exacerbated by fluctuations in the price of oil and gas and by increased liability for environmental damages. Oil exploration is certainly no place for the small or faint of heart, and the odds in other minerals could not be much better.

If significant risk aversion is present, it will drive small firms out of the industry or reduce their level of participation. Small firms cannot afford to gamble on expensive exploration programmes requiring offshore drilling platforms or Arctic operations. They cannot spread and diversify explorations risks as big firms can.[24] Every firm subtracts a risk discount from its bid for a mineral lease; the smaller the firm, the bigger the discount. In addition, the exit of the smaller firms would reduce the number of bidders, so that revenues would be reduced via the relationship between the number of bidders and the size of bids (see my other article "An Economic Theory of Mineral Leasing," in this volume).

By performing the early geophysical and drilling work in virgin areas, the government could reduce the risk to firms by decreasing the size of the outlay required and increasing the probability of success on a given venture.

Outlays would be smaller because the exploratory work would have been performed, and large tracts would not be required in order to capture information spillovers and utilize scale economies. The reduction in risk would encourage the entry of smaller firms and increase the government's revenues on lease sales. The expected effect on the size of firms was demonstrated at Prudhoe Bay, Alaska, where the small firms purchased leases mostly after the big, risky wildcats had been successfully completed in 1967-68.[25] Lease sale statistics anywhere should reveal that small firms take the small, low risk opportunities and large firms explore the big, risky prospects in virgin areas.[26] Lease bids would rise on average, but some tracts would receive lower bids because of unfavourable information, such as dry holes. Government disclosure of all information would be mandatory. If information were withheld or edited, the government would lose credibility, and its exploration programme would lose much of its value.

Industry Concentration

If information spillovers, scale economies, and risk aversion were significant problems, mineral exploration activities would be dominated by a few large firms, but McKie[27] and the big oil companies[28] argue that mineral exploration is not concentrated, that there are no significant barriers to entry, and that the small companies make most of the discoveries. There are thousands of firms in the prospecting and wildcatting business, some of them operating with very little capital or specialized knowledge, and independent oil operators make over three-fourths of the new field discoveries in the United States.[29]

Data published on the number and importance of small exploration firms are deceiving. It is true that petroleum refining, for instance, has a low concentration ratio compared with some other industries and that petroleum exploration has an even lower ratio; but these ratios do not tell the whole story. The concentration of power in the whole petroleum industry may be much higher because of interlocking directorships, agreements to share technology, and joint ventures. Ridgeway studied affiliations of directors, acquisitions of coal, gas, and uranium reserves, and interchanges of refining technology, and he concluded that oil companies are acting in concert to monopolize the last of the world's energy resources.[30] His claim is exaggerated in my opinion, but the petroleum industry as well as other mineral industries would appear to be more concentrated than is indicated by the size distribution of firms.

Information spillovers, scale economies, and risk aversion may force oil companies into cooperative exploratory ventures that would put the small companies in the back seat, so to speak. A small company may be drilling a given well, but a big company possibly owns most of the leases and has a significant interest in the well.[31] The statistics on who is discovering oil are deceptive and should be changed so that they credit a proportionate part of the oil discovered to the interest holders rather than just to the operator of record.

The little firms' roles are deceptive in another way. The size distribution of oil fields is highly skewed, with most fields being very small. The largest 200 out of 10,000 fields in the United States account for more than half of the reserves.[32] There is strong evidence that small, independent companies explore the smaller oil structures. The big ones that remain untested are in the Arctic, offshore, or very deep, where the independents cannot afford to go. McKie, Shearer, and the industry people with whom I have spoken agree on this point.[33] Although public data on the amount of oil discovered by firms of different sizes were not found, the United States Bureau of the Census has published some illuminating figures. Of the thousands of oil companies operating in 1963, the biggest eight made 45 per cent of the exploration expenditures. Unless the big companies were incompetent, they found as many barrels per dollar as the little guys. The average well drilled by the biggest eight was considerably deeper and presumably tested a larger geological structure than the wells drilled by the others.[34] In short, the independent companies discover three-fourths of the new oil fields in the United States, but they could not be discovering three-fourths of the oil.

If the oil and gas exploration business is concentrated, the exploration for other types of minerals must be even more concentrated. The metals industries, for instance, have higher concentration ratios than oil and gas. Just a few firms dominated the production of some metals, and concentration in the whole production process suggests concentration in the exploration step. And world cartels like CIPEC, for copper, and IBA, for bauxite, must affect exploration activities.

To summarize: arguments that mineral exploration is not concentrated appear insufficient to negate the market-failure arguments for government intervention presented here. Just as the government performs and subsidizes basic research because of information spillovers, scale economies, and risk aversion, the government should perform or supervise the early exploratory work on its own mineral properties. This would include geophysical surveys and a certain amount of digging and drilling. It would require further analysis to tell how far through the exploration stage and into the development stage the government should go. Government exploration could probably go considerably beyond its present level in the United States and Canada with benefits exceeding costs. By performing additional exploration and publishing the results, the government could prevent the wasteful duplication of exploratory activites, encourage the participation of smaller firms, accelerate the rate of exploration, and increase its mineral revenues.

Notes

1. Minister of Mines and Petroleum Resources, Province of British Columbia, *Annual Report, 1972,* Victoria, B.C.: British Columbia Department of Mines and Petroleum Resources, 1973, pp. A64-A77.

2. "Federal Development of Reserves Urged," *Oil and Gas Journal* 71 (2 April 1973): 38-39.

3. Lee E. Preston, *Exploration for Non-Ferrous Metals* (Washington, D.C.: Resources for the Future, 1960), pp. 130-41.

4. Joseph Stanley Clarke, *The Oil Century* (Norman, Oklahoma: University of Oklahoma Press, 1958), pp. 10-35.

5. Edward Miller, "Some Implications of Land Ownership Patterns for Petroleum Policy," mimeographed, November 1972, forthcoming in *Land Economics.*

6. James W. McKie, "Market Structure and Uncertainty in Oil and Gas Exploration," *Quarterly Journal of Economics* 74 (November 1960): 566.

7. Frederick M. Peterson. "Two Externalities in Petroleum Exploration," in Gerard M. Brannon, ed., *Studies in Energy Taxation,* (Cambridge, Mass.: Ballinger Publishing Co., 1975), pp. 101-13.

8. M.S. Hedley, *The Mineral Industry in British Columbia* (Victoria, B.C.: British Columbia Department of Mines and Petroleum Resources, 1966).

9. R.H. Coase, "The Problem of Social Cost," *Journal of Law and Economics* 3 (October 1960): 1-44 and McKie, "Market Structure": 566-68.

10. Miller, "Land Ownership." As an indication of the transaction costs, it took the Humble Oil Company seven years to lease all the land over a possible field in East Texas. The field was a success, and they were lucky to keep it a secret that long. See McKie, "Market Structure," p. 566.

11. McKie, "Market Structure", p. 565.

12. "Mining Developments in British Columbia Continue at a Brisk Pace," *Engineering and Mining Journal* 171 (September 1970): 100; Minister of Mines and Petroleum Resources, *Annual Report, 1972.* pp. A70-A74.

13. Preston, *Non-Ferrous Metals,* p. 32. Louis B. Slichter, "Geophysics Applied to Prospecting for Ores," in Alan M. Bateman, *Economic Geology, Fiftieth Anniversary Volume,* Part II (Lancaster, Pa.: Economic Geology Publishing Co., 1955), pp. 885-86.

14. Louis B. Slichter, "Some Aspects, Mainly Geophysical, of Mineral Exploration," in Martin R. Hubery and Warren L. Flock, eds., *Natural Resources* (New York: McGraw-Hill, 1959), pp. 376-77. Mineral Claims are subsequently patented in order to obtain ownership rights. Part of the decline in patents may be due to factors other than mineral depletion such as reduced legal advantages of converting claims into patents.

15. U.S. National Science Foundation, *Report of the Advisory Committee on Minerals Research to the National Science Foundation* (Washington D.C.: National Science Foundation, 1956), p.3.

16. Clark, *Oil Century,* pp. 10-28.

17. Slichter, "Some Aspects," pp. 376-7.

18. Ronald A. Shearer, "Nationality, Size of Firm, and Exploration for Petroleum in Western Canada 1946-1954," *Canadian Journal of Economics and Political Science* 30 (May 1964): 211-27.

158   *Frederick M. Peterson*

19. Richard A. Musgrave, *The Theory of Public Finance* (New York: McGraw-Hill, 1959), pp. 137-39.

20. Hayne E. Leland, Richard B. Norgaard, and Scott R. Pearson, "An Economic Analysis of Alternative Outer Continental Shelf Petroleum Leasing Policies," mimeographed, University of California, Berkeley, September 1974.

21. Milton Friedman and L.J. Savage, "The Utility Analysis of Choices Involving Risk," *Journal of Political Economy* 56 (August 1948): 279-304.

22. C. Jackson Grayson, Jr., *Decisions Under Uncertainty* (Boston: Harvard Business School, 1969), p. 10.

23. The theory of the second-best is relevant here. See Kelvin Lancaster and R.G. Lipsey, "The General Theory of the Second-Best," *Review of Economic Studies* 24 (January 1957): 11-32.

24. Maurice Allais, "Method of Appraising Economic Prospects of Mining Exploration over Large Territories: Algerian Sahara Case Study," *Management Science* 3 (July 1957): 285-347; Louis B. Slichter, "Mining Geophysics," *Mining Congress Journal* 45 (May 1959): 38-39.

25. "Smaller Firms Score Big in Huge North Slope Sales," *Oil and Gas Journal* 67 (September 1969): 23-25.

26. McKie, "Market Structure," pp. 548-64.

27. McKie, "Market Structure."

28. M.A. Wright, "Look at the Facts," *Exxon U.S.A.,* First Quarter, 1974, pp. 6-9.

29. "Independents Widen Lead in U.S. Oil and Gas Search," *Oil and Gas Journal* 72 (1 July 1974): 11-15.

30. James Ridgeway, *The Last Play* (New York: E.P. Dutton, 1973).

31. McKie, "Market Structure."

32. M.A. Adelman, "Economics of Exploration for Petroleum and Other Minerals," *Geoexploration* 8 (3 December 1970): 143.

33. McKie, "Market Structures." Shearer, "Western Canada."

34. U.S. Bureau of the Census, *1963 Census of Mineral Industries*, 1, "Summary and Industry Statistics" (Washington, D.C.: Government Printing Office, 1967), pp. 13B-99. The larger structures would have been tested already if they were not deeper or in some other way more expensive to test.

Comment

RUSSELL S. UHLER

In order to determine the impact of alternative leasing policies on mineral industry exploration and development decisions and on government revenues, it is important to understand the effects of two pervasive conditions which Frederick Peterson has discussed; namely, risk (uncertainty) and the lack of markets for the sale of information. Each condition is believed to lead to an underallocation of resources relative to that allocation which is socially optimal.

RISK

If decision markers are risk averse as is commonly supposed; if, in other words, their marginal utility of additional return (income) is diminishing, then they will always prefer a certain return of a given amount to a probabilistic one with the same expected (mean) amount. This means that there exists a certain return which is less than the mean value of the probabilstic one, yet between the two the decision maker is indifferent. The difference between this certain and uncertain return may be thought of as the insurance premium which the decision maker is willing to pay for certainty. Even though a decision maker may want to rid himself of this risk by buying insurance against it and thus shifting it to others, in most instances it is not possible to do so. This is certainly true in mineral exploration and development, as it is for most business operations, because the risk is uninsurable under the general provision of "moral hazard." This provision comes into effect whenever the act of insuring a risk may significantly change incentives and, thus, the underlying probabilities upon which the insurance is based, and it is the main reason why insurance exists for relatively few categories of risk in our economy.

These propositions regarding risk averse behaviour are obviously important to mineral leasing decisions in, for example, a bonus bidding lease arrangement. Under conditions of complete certainty it is easily shown that bonus bidding is the preferred method of leasing. The most efficient firm or syndicate of firms will win the bid, and, given sufficient competition, its profits will not be excessive and government revenues will be maximized. Moreover, bonus bidding does not distort decisions on the life of the deposit as do royalties, for example. But under conditions of uncertainty the propositions stated earlier lead to the conclusion that bonus bidding, even if a large number of firms are competing for the lease, will reflect the risk averse behaviour of firms and will result in a winning bid which is lower than if the firms were making decisions on an expected value of return basis (that is, if they were risk neutral). Thus government revenues from pure bonus bidding are less than they should be. Furthermore, it is easily shown that as the level of uncertainty increases so does the amount of risk aversion.

It has been effectively argued in the literature on the economics of uncertainty that socially optimal resource allocation could be obtained if risk neutral behaviour could be assured. But such behaviour cannot occur without appropriate insurance markets, and such insurance will not be provided because of moral hazard. Thus, a *first best* or risk neutral world cannot be expected to be achieved through the provision of insurance for these risks, either by the private insurance market or by government. Obviously, government attempts to provide such insurance would also be confronted by the problem of moral hazard.

But insurance is not the only institution in the private sector of the economy for shifting risk, the other major one being common stock ownership. Large firms or syndicates of firms with many small stockholders may be able to shift risk relatively well, which, one might expect, would reduce the risk averse behaviour of decision makers. Moreover, in addition to being able to shift risk, such firms are able to pool it through diverse operations. Nevertheless, even if one accepts the argument that large firms have the ability to shift risk relatively well, it is still likely that decision makers within the firm will act in a risk averse manner. For the most part these decision makers are professional managers, and large losses, even though they may not injure the individual stockholder significantly, may be taken as a sign of incompetence leading to management reorganization and job losses. These arguments suggest that although institutions exist within the private economic system for shifting risk, appropriate insurance markets do not and will never exist for complete shifting of business operating risks, and extensive common stock ownership probably does not prevent risk averse behaviour.

Given that moral hazard is the primary reason why private insurance markets fail to provide protection against business operating losses, a *second best* alternative to complete insurance coverage is partial insurance or coinsurance. Although the provision of this type of insurance ought to lead to greater neutrality in the behaviour toward risk, its use has been limited. Nevertheless, the concept of coinsurance suggests how mineral lands may be leased to promote more risk neutral behavior. Specifically, a lease programme whereby government shares in both profits and losses is formally similar to coinsurance. Part, but not all, of the risk is shifted to the government, but the firm's incentive to be efficient remains because only part of the losses are covered by the government, and only part of its profits are taxed. As long as expected returns to mineral exploration and development in a region are positive, the government will, in the long run, receive positive revenues from profit and loss sharing. There are, of course, certain difficulties with administering such a leasing policy. Profits and the method of determining who gets the exploration and development rights must be defined. However, these difficulties do not appear to be insurmountable. Although I do not present a detailed theoretical treatment showing the advantages of profit and loss sharing over other leasing policies here, these introductory points suggest that no other policy will do better in promoting risk neutral behaviour.

I have suggested that we cannot expect risk neutral resource allocation to result from private decision making in any sector of the economy, including the mineral industry. Is it possible to achieve these results in another way? Peterson and others have argued that of all economic units the government has the greatest capacity to pool and shift risk and, hence, is a logical candidate for involvement in mineral exploration. Indeed, the profit sharing plan is a straightforward attempt to get government to share some of the risk. But why not go all the way and let it carry all of the risk, not by insuring private firms, which we have already seen will not work, but by doing the exploration and development on its own? Although I am not inherently opposed to complete government control in this area of economic activity, I do have reservations. First of all, I am not sure that a crown corporation would be more risk neutral than a large firm or syndicate. Both have the capability of shifting risk, one to common stockholders and the other to the people of the province (or nation). But it was argued earlier that, even with this risk shifting ability, private decision makers will be risk averse, because if large losses should occur, their jobs would be in jeopardy. I think the same is likely to be true of decision makers in crown corporations. Losses can embarrass the government and possibly even help lead to its downfall at the hands of the electorate, so that preferences for job security by crown corporation management may also lead to risk averse behaviour. Indeed, it may be that government corporation managers are more risk averse than their private counterparts. With these arguments in mind, I would suggest that the case for government monopoly in mineral exploration based on attitutes toward risk alone is not very strong.

MARKETS FOR INFORMATION

Besides risk, a second condition existing in mineral exploration and development which influences resource allocation is the lack of markets for information. Peterson refers to this condition as "information spillover." The problem is the classical one of externalities. Firms will tend to invest suboptimally in exploration from the viewpoint of society as a whole, because the marginal private benefits of information from additional exploration effort are less than the marginal social benefits. This is because information is so easily and cheaply transmitted that it is difficult for its discoverer to completely capture, or internalize, its social value. This condition will lead firms to hold back investment in exploration as they wait for other firms to provide full valuable information to them. If this behaviour is widespread, it leads to underinvestment by the industry as a whole. Since markets for information can never be expected to exist in our economy, all possibility of internalizing its value is ruled out; so if firms are to invest optimally, some other way must be found which allows them to reap fully the returns of their investment in information gathering. One such possibility in mineral exploration is simply to make land tracts large enough that the spillover effect is reduced. Another possibility is to eliminate the problem altogether by having a government enterprise (crown corporation) do the exploration. A government monopoly (like a private monopoly, for that matter) would not be faced with the information spillover condition and, hence, would behave as if it did not exist. This condition undoubtedly provides the strongest case for complete government involvement in mineral exploration. Although we have qualitative evidence that social gains are possible from such action, we have no idea what the magnitude of these gains would be. If they are small then government takeover is simply not worthwhile. We need empirical studies of the benefits and costs of such action before launching upon such a new policy.

The Role of Public Enterprise

My remarks are, firstly, about the role of governmental enterprise generally and, secondly, about some of the considerations involved in using governmental enterprise to foster greater control by the citizens of Canada and British Columbia over their own mineral industries. My view is that governmental ownership of producing operations is not generally the most effective way of accomplishing the social ends for which it is currently being advocated in these industries. Nevertheless, I have a few suggestions how some of the major disadvantages of public enterprise with respect to efficiency and responsibility might be overcome.

Government owned enterprises in the English-speaking countries have seldom owed their existence to an anticapitalist ideology. It is, in fact, hard to detect any systematic difference in motive, organization, or operation between the national, state and provincial, or municipal enterprises established during the incumbency of labour, socialist, and agrarian radical parties and those implemented by Tories of various names and complexions.

Despite the vast amount of existing governmental enterprise today in capitalist countries, and despite the importance of socialist movements and socialist thought in the history of modern civilization, the scholarly literature on public enterprise is remarkably skimpy. Rigorous comparisons —theoretical or empirical—of the economic performance of governmental and private enterprises in the same industry are, to the extent I can determine, nonexistent.

GOVERNMENT OWNERSHIP

The case for government ownership of undeveloped land and natural resource stocks rests on a broader base than that for government ownership of producing enterprises. The intrinsic value of any resource in its natural state is the difference between the value of goods that can be produced from it and the cost (in terms of labour, capital, materials, and organization) required to produce those goods. The size of this residual is not the product of any person's labour or enterprise; most of the economic value of an *in situ* resource and its appreciation over time result from such diffuse causes as the increase in population, the general advance of technology, the decline in real transport costs, or directly from governmental outlays on roads or geological mapping. On these grounds, it has become almost an axiom of distributive justice (however commonly violated) that the intrinsic value of natural resources should not be privately appropriated.

Other classical grounds for government ownership of natural resources are the desire to control external costs or capture external benefits of their exploitation, and the expected divergence of private capital costs from the social rate of time preference, which is said to result in too rapid (or too slow) development of the resource. I am skeptical about the universal applicability of the last of these arguments; who, indeed, knows what society's true discount rate should be, and why are politicians and civil servants expected to be more sensitive to it than to entrepreneurs? This reservation notwithstanding, I believe that a presumption in favour of government ownership of undeveloped land and resources is generally justified.

Turning to productive enterprise, however, there are three main economic rationales for government ownership in a capitalist society. First is the use of the state to establish or maintain productive activity that would not be profitable as private enterprise, but whose external benefits are deemed to justify a subsidy out of the public exchequer. A subsidy does not, of course, require state ownership, because either private or governmental enterprise could enjoy that subsidy. In either case, support could take the direct form of providing capital or operating expenses from the Treasury or the indirect form of tax exemptions and the use of public resources at less than their cost of fair market value. State ownership, however, may well make a subsidy more palatable to the public, because it does not conspicuously enrich (or appear to enrich) a few private entrepreneurs.

Within the category of public ownership as a vehicle for subsidization are the numerous instances of private enterprise socialized because of chronic insolvency or imminent liquidation, including the Canadian National Railways, most of the British Labour Party's nationalizations after World War II, and the recent takeover of rail passenger transportation by the United States government.

In other cases, the motive for government ownership has been the creation of "public goods," products (or by products) of an enterprise whose value a private owner could not expect to recover by market pricing. Examples of such externalities are flood control by hydropower projects and the promotion of literacy and national unification by the postal system.

Military necessity has been another justification for producing goods in state enterprises which might not meet the test of the private market.

Nineteenth-century America had government lead mines and arsenals and plantations for naval stores; the processing of nuclear fuels now remains a governmental activity on security grounds. Many public transportation and communications ventures were begun as defence projects in Alaska, the Yukon, and British Columbia: examples are the ALCAN highway and the White Alice communications system.

Second among the rationales for public enterprise is the perceived inability of private business, because of the great size or risk of the venture in question, to assemble sufficient capital. This tradition in North America began with state ventures in canal and rail development in the early nineteenth century, then extended to river control and irrigation projects, and continues into the present in enterprises like COMSAT and Panarctic Oils. In many of these cases, the proposed activity was expected to be self-sufficient in the long run, on the basis of the revenues from its product or service, but state initiative was seen as necessary to take advantage of scale economies or to overcome high risk threshholds.

The third circumstance seen to justify government ownership is possession by an enterprise of monopoly power and/or exceptionally rich natural resources, either of which can produce substantial "unearned profits" or rents. Government ownership is one means either of preventing monopoly exploitation of consumers (or monopsony exploitation of workers and sellers) or of collecting for the public treasury monopoly profits or resource rents that would otherwise be captured by the private owners.

Government takeover of profitable businesses has been rare in the English-speaking world. There have been a few instances of ideologically motivated nationalization, but it is instructive to note that these have often been reversed, as in the cases of the iron and steel industry in Britain and, more recently, the grain trade in India. The remaining cases have principally been those of utilities—grain elevators, street railways, water, electrical or telephone systems—which had a monopoly ("natural" or otherwise) in a local service area. In the last category it is often hard to distinguish between the instances where government took over to prevent private exploitation of monopoly power and those in which government saw a monopoly as an opportunity to exploit an assured source of revenue for itself.

There are, of course, a variety of cases which overlap two or all three of these categories. Economic development of a poor or sparsely settled region is often advanced as a justification for public enterprise in transport, communications, or electrical power. In these instances the premise is often that the region lacks capital or capital markets and only the state can mobilize resources on the desired scale. At the same time, the project is seen to encourage growth by its ability to widen markets or otherwise cut costs for commodity producing sectors of the regional economy. Once established, moreover, such an enterprise may have a monopoly status, with the power to abuse or exploit that status, and seem thereby to demand public control or ownership.

Many governmental enterprises (and regulated utilities, which they resemble in important respects) combine subsidies for some activities with appropriation of monopoly rents or resource rents from others. A common practice in both regulated private firms and government enterprise in transportation, communications, and utilities is *cross-subsidization*, in which monopoly profits earned from one area, line of business, or class of customers are dissipated in subsidizing others that are deemed to be socially meritorious. Thus, airline and railroad tariffs on heavily travelled route segments typically exceed cost (including a "fair" return), while service on lower density segments is provided at a loss. Hydroprojects in the Western United States typically subsidize users of irrigation water from revenues earned by water sales to municipalities and industry and by sales of electric power. In Alaska, revenues from both state and federal timber sales are sacrificed to support otherwise uneconomic lumber and pulp mills.

Turning to the mineral industries of Western Canada, there is little evidence that suggests they need to be subsidized by formation of a public corporation or otherwise. The province of British Columbia does not have, for example, a great but decaying industry upon which the community depends both for energy and employment, as the British had in the coal mines of the 1940's.

The British Columbia-Yukon Chamber of Mines may occasionally assert that each mining job generates seven additional jobs in supporting industries, but there is no respectable analytical foundation for such a claim. Even if the extractive industries had such an employment multiplier, it does not necessarily follow that job creation *per se* is a benefit that deserves subsidization from the public purse, much less the creation of a government enterprise. New employment opportunities are a *net* benefit to the existing community only to the extent that they are filled by residents who would otherwise be unemployed or working at more poorly paid jobs. In an "open economy" like that of British Columbia or Alaska, there is no predictable relationship between local job creation and local unemployment, because new employment opportunity attracts immigrants who tend to offset the employment gain. Even if the new jobs directly created were reserved for long-time residents, displacement of residents from old to new jobs and their replacement by nonresidents can be expected to make overall unemployment rates relatively unresponsive to employment growth.[1]

National self-sufficiency in minerals and the earning or retention of foreign exchange are sometimes claimed as external benefits of mining that justify preferential treatment. In Canada, paradoxically, some of those (including the National Energy Board) who place the highest priority on national self-sufficiency in one or another mineral resource tend to advance policies that *deter* investment, on the grounds that the beneficiary of current development tends to be the export market, at the cost of future diminished Canadian self-reliance. In this country, moreover, balance of payment effects are often used as part of a case *against* mineral development for export rather than in favour of it. In a world of floating exchange rates, however, one might question whether there is any relevance at all to the balance of payments problem in its usual sense. Finally, the impacts upon environmental quality and the dispersion of population are more likely to be regarded as external *costs* of mining than as *benefits* that justify the government's promotion of mining ventures that otherwise would not be self-supporting.

It is hard to make a respectable case that mining (including oil and gas production) creates beneficial *externalities* for the surrounding community, as distinct from the net value of the minerals produced, or the factor payments (wages, profit, rents, and taxes) which make up that value. The current Canadian interest in state enterprise in the mineral industries does not seem, in summary, to be a result of the belief that they are inevitably unprofitable under private enterprise. On the contrary, it rests in part upon the notion that mineral extraction can indeed be very profitable, and that unearned profits (rents) ought to be controlled and disbursed in socially approved ways. This attitude is sometimes experienced as a concern whether the people of the nation or of the province, who are the nominal owners of its natural resources, are receiving as high a return for the products of their land as they might. One issue is, in short, whether the state is effectively maximizing its revenues from disposal of minerals.

Where effective capital markets exist together with a large enough number of potential operators to create workable competition for resource rights, the government (as landlord and/or sovereign) is more likely to maximize its revenues if it does *not* engage in production. This conclusion does not presume that particular government owned entities are necessarily less enterprising and less effective in cost control than profit motivated private corporations. There may well be a bias against efficiency in most forms of state enterprise, if only because their owners (the public) and managers do not have a clearly defined standard of performance as private managers have in the imperative to maximize the present worth of their firms. But more importantly, by operating a productive enterprise in the extractive industries, the government loses the ability it would otherwise have as landowner to exploit the competition among potential private operators.

At oil and gas lease auctions in the United States, for example, the bid prices on a single tract may vary by a factor of two, ten, or even one hundred. These variations reflect widely differing geological evaluations of the tract, exploration strategies, and capital and other costs. Thus each tract tends to be won by the bidder with the *most favourable combination of capital cost and expectations* among all the bidders regarding future product prices, the particular tract's recoverable reserves, and their development and lifting costs. The landlord (state or private) who operates on his own land, however, would have only one management team, one exploration strategy, one team of geologists and engineers, and one supply function for capital. Only by rare accident would the landlord's *actual* performance over the average of all his properties tend to be better than the *expectations* of the most optimistic bidder. If, therefore, he were to lease each tract to the highest bidder among the competing operators, he might anticipate receiving a greater net revenue on each property than he could expect from developing the property himself.

The foregoing prediction is implicitly supported by empirical studies of United States Outer Continental Shelf leasing by Walter Mead and others, who show that successful bidders on the average earn a discounted cash flow rate of return on lease acquisition costs substantially less than the oil industry's average rate of return on capital (see Mead's article, "Cash Bonus Bidding for Mineral Resources," contained in this volume).

The effective use of competition to optimize revenues does not dictate the use of a cash bonus bidding system for *all* minerals or even for petroleum under every circumstance. The degree of knowledge or uncertainty regarding the volume and value of minerals present and their cost of extraction, the relative weight of fixed and variable costs in total extraction costs, the number of potential competitors, and the relative preferences of the government and private operators for certain present income versus uncertain future income are all appropriate considerations in the choice of leasing or disposal systems and taxes on the mineral industries. These questions have been discussed elsewhere,[2] and other articles in this volume give close attention to the relative merits of location, leasing, and sale as systems for disposing of minerals; to royalties and severance taxes, and whether they should be reckoned on gross value or net profits; to bidding on cash bonuses, deferred bonuses, gross or net royalty rates; to the use of acreage rentals; to the optimum size and configuration of tracts; to the amount of geological information the landlord ought to obtain and publish before opening land for lease or disposal; and to the duration of the primary term of a lease or permit, its terms for renewal, and so on.

In summary, state enterprise in the business of developing and producing minerals is surely *one* way to capture and redistribute resource rents, but it is unlikely to be as effective a device for maximizing those rents as the combination of a leasing system that takes full advantage of competition

among private firms (considering the technology and institutional characteristics of each branch of the mining industry) and an appropriate tax
system.

The most powerful cases for public enterprise in developing regions (like
much of Western Canada and the Territories) relate to transport facilities
which create external economies for other economic sectors, including
mining. In these instances, both the first motive for socialization (the desire
to subsidize) and the second (the need to overcome barriers of scale and
risk) may justify investment by the government on projects into which
private enterprise will not venture. Neither of these motives, however,
creates a case for state enterprise in mineral extraction. Capital and
enterprise for mineral exploration and development are plentiful and
mobile. Specialized technical inputs, such as geophysical surveys, drilling,
and heavy construction can be purchased on contract in a highly competitive market (so that great petroleum and mining companies carry out very
few of these activities themselves). Capital sums in the hundreds of millions,
or even billions of dollars can be mobilized privately, even without
government guarantees, for projects like the Trans-Alaska pipeline, in
remote regions.

Development of minerals, like the collection of revenues from their
development, is likely to be more rapid and more efficient if it utilizes the
diversity of skills, techniques, enterprise, and access to capital in the private
economy and the competition between firms differently endowed in these
respects. Nationalization or municipalization of producing operations, in
my view, has an inevitable price—both in state revenues and in social
efficiency—the payment of which must be justified on other grounds.

The hard core of Canadian interest in public enterprise today seems to
stem not from a perceived shortage of capital and entrepreneurship (much
less an ideological opposition to capitalistic enterprise as such), but from a
perceived surplus of foreign capital and entrepreneurship. The problem, it
seems, is to assure that mineral development (and, presumptively, related
activities like oil refining or oil and gas transportation) are under the control
of Canadians or the people of British Columbia, rather than great
multinational (read United States) corporations.

It is probably not politic of me as an American to ask what practical
difference the nationality of a company's owners or management makes as
long as it is subject to the same laws (and obeys them) and pays its proper
share of taxes. Foreign companies in Canada have often been berated for
not paying enough taxes, but it was after all a *Canadian* decision, reflecting
a long-standing Canadian developmental philosophy that the extractive
industries should remain largely untaxed. The satisfactions and grievances
of Americans regarding the major oil companies apply in the same way to

Mobil and Texaco, which are domestically controlled, as they do to Shell and Sohio, which are foreign controlled. I am afraid I don't see how any more in the way of real resources for Canadians could be squeezed from a government owned business than could be squeezed from American or Canadian owned private enterprise under a well designed leasing and tax system. Nevertheless, one billion American dollars invested in Canada is more conspicuous than one billion Canadian dollars invested in the United States, and the nationality of your managers and stockholders obviously does make a difference to many Canadians. (I must confess, also, that some Americans become hysterical about the very idea of the Arabs or Persians taking over United States businesses.)

Government enterprise is one way to "nationalize" the mineral industries, but it is not, of course, the *only* possible way to foster Canadian equity and enterprise. Stricter nationality criteria could well be applied to holders of claims and leases or of permits to build pipelines, concentrating plants, and refineries. Such policies raise the further question, however, whether there is in fact enough private equity capital and enterprise in Canada to effectively take the place of foreign equity and enterprise. This is an empirical question to which I do not have an answer. If the answer is negative, consideration must be given to the fact that establishing a government enterprise does not *create* any new Canadian resources. It only uses tax money or potential resource revenues to bid capital and talent away from some other employment in Canada. The cost of Canadianization (either by restrictive licensing of private industry or by government enterprise) may be minimal, however, if preferences for nationals result in bidding home significant amounts of Canadian capital and Canadian talent which would otherwise be employed in other countries. (Presumably, the net effects of even these moves would have to take into consideration remittances that would otherwise flow back to Canada from investments abroad.)

Government capital need not be regarded strictly as a *substitute* for private capital, Canadian or foreign. In North America during the first half of the nineteenth century, and in almost every country at one time or another, state companies were used as a vehicle to *attract* foreign debt or equity capital, usually British, to ventures they would not otherwise consider. A government owned (or guaranteed) railroad company was often naively regarded by Lombard Street as a safe investment, while the promotions of unknown overseas entrepreneurs were viewed with little regard in the world's principal money market.

The use of government participation is still a major instrument for encouraging foreign investment in developing areas. Joint ventures between American, European, or Japanese private companies and governmental entities of the host country are common in almost all the extractive industries and in many countries at different levels of economic development. Canada has at least one government enterprise created largely with this function in mind, Quebec's SOQUEM, whose activity consists mainly of joint ventures in mineral exploration with private companies. Petro Canada also *seems* to be interested in this kind of approach.

In addition to being a means by which domestic enterprise becomes a trustworthy borrower (or partner) of foreign capital, state enterprise can also be a means of offsetting a shortage of domestic equity and entrepreneurship. In this role, it has one advantage over promotion of domestic private enterprise through nationality restrictions on investment, management or licences: it avoids the spectre of open discrimination, which could lead to retaliation and might otherwise undermine trade and investment relationships that are beneficial to Canada. The nearly open border allows this country to draw on a much larger pool of capital, technology, and talent than it would with policies fostering autarky. Although this openness is a major element in the ambiguity and insecurity of Canada's national identity, its economic benefits to Canada are relatively greater than they are to the United States. (That is, its impact on the size of the resource pool available to Canada is greater than on the size of the pool available to the United States.) It is therefore a circumstance to be modified only carefully and selectively. Establishing a provincial oil company is one way of containing the side effects of a move in the direction of autarky in a single industry. Such a move might, in fact, limit these side effects even in the industry in question. Suppose the best candidate for executive officer for a British Columbia based oil company were a Texan; there might well be fewer misgivings about hiring him to work for the province than about his heading a subsidiary of an American private firm.

I will conclude this article with some suggestions for the structure and policy of public corporations in the mineral industry, suggestions aimed at combining some of the best features of government and private enterprise, rather than their worst.

First, before establishing a governmental enterprise, be clear what its purpose is to be, what the incentive for the management to accomplish that purpose will be, and, quite rigorously, what will be the measure of the enterprise's success. (I owe this first and most vital point to Milton Moore's critique of the draft of this article.)

Second, do not set up a monopoly. There is no surer formula for inefficiency and social irresponsibility. Economies of scale do exist in mining and petroleum exploration, but they are very small when compared to some other industries or relative to the opportunities for development in an area the size of British Columbia. In petroleum refining, the minimum efficient size of a refinery is probably about the size of the British Columbia market for petroleum products, but if a new government owned refinery needs a monopoly or protectionist legislation to be profitable, it will almost certainly be a serious burden on consumers. Industries in which scale economies are narrow and where ingenuity and intuition are still crucial, as in mineral exploration or onshore oil and gas production, are probably not the most appropriate candidates for nationalization; but where it is determined to establish a state enterprise, consideration might be given to the establishment of more than one competing public enterprise.

Third, do not clothe the corporation in sovereign immunities. Such immunity can be, and often is, a cover for inefficiency, irresponsibility, and even lawlessness. The corporation should be suable; it should pay taxes or their equivalent (federal, provincial, and local); and it should be subject to environmental and safety laws and regulations and, above all, subject to the bankruptcy laws. Its operations should not be protected by any version of an official secrets act. There is no good reason why the directors, officers, and employees should be excused from the same civil and criminal liability for their actions to which their counterparts in private enterprise are subject.

I would urge hesitation even in providing guarantees for the corporation's debt. A public mining or oil corporation will be pursuing a line of business in which private enterprise regularly borrows money without such guarantees. The more intense scrutiny of bankers and underwriters toward a corporation whose debt must stand on its own merit might well save the corporation's owners—the public—more money than the small interest differential associated with government guarantees.

Fourth, give the public and the corporation's officers and staff a material interest in its success and its efficiency. The government need not hold all the shares but only a controlling interest, not necessarily even a majority. One block of shares (enough to elect at least one director) can be held in trust for the company's employees and voted by them. The remainder of the shares would be offered to the public; they would be voted by their owners and publicly traded. Not only would this provision broaden interest and participation in management, but the market price of publicly traded shares would be a continuing indicator of management performance and of the value of the government's equity. I see no compelling reason to restrict share ownership to residents; it might in fact be useful to encourage minority participation by major oil companies or mining companies. A residence requirement for shareholders, however, would reinforce symbolically the corporation's identity as a national or provincial instrument, and would, of course, limit remittance of dividends abroad.

Fifth, the corporation's policies should be responsive to public policy but not bend to every political wind. I would suggest that only a minority of the government directors serve at the pleasure of the Cabinet and be regarded as spokesmen for its policies. The remaining directors representing the government's equity would be chosen indirectly for long and staggered terms.

Sixth, the corporation should be under pressure to pay dividends. A majority of the shares (and directors) should represent parties who have a material interest that the corporation *not* retain, reinvest, or dissipate all its earnings: private shareholders, the employees, and the members who serve at the pleasure of the Cabinet (who would presumably be responsive to the fiscal interest of the government). The influence of this group will be a constant corrective to tendencies of management, inside directors, and permanent directors toward complacency, empire building, pyramid building, or gold plating.

Seventh, maintain a clear distinction between the corporation and the government as landowner. The public enterprise should obtain resource rights on crown lands only in competition with other prospective operators. The corporation should not receive a concealed (and indeterminate) subsidy by access to resources at no charge or at a lower price than a competitor might offer. If it must have a preferent right, let it be at most a right to match the highest bidder.

A preferent right on the best offshore leases is a feature of the federal oil and gas corporation (FOGCO), proposed recently in the United States Congress. In view of the prices oil companies have been recently willing to spend in these lease sales, such a preference would guarantee that FOGCO would appear profitable, however incompetent its management, and that the federal treasury would lose billions of dollars in lease revenues.

Eighth, take advantage of the division of labour and competition. The corporation should not attempt to do for itself the things that even the greatest oil and mining companies contract out to others, such as seismic surveying, core drilling, well drilling, well logging, and construction. There is virtually no chance that a state corporation could improve on the performance of private firms in these exceedingly competitive areas.

In summary, I am generally skeptical of the case for public enterprise in the minerals industry but hopeful that such enterprises could be established free of many of their usual shortcomings, providing some thought is given to their purpose, organization, and standards of performance.

Notes

1. In a study aimed at projecting the employment impact of the Trans-Alaska pipeline, we found that *unemployment* in individual labour market areas was almost totally insensitive to the level of *employment*; that is, on a *net* basis, at least, new jobs in Alaska's petroleum and wood products industries and government were entirely filled by immigrants. [Arlon R. Tussing; George W. Rogers; and Victor Fischer; with Richard Norgaard, and Gregg Erickson, *The Alaska Pipeline Report: Alaska's Economy and Gas Industry Development and Impact of Building and Operating the Trans-Alaska Pipeline*, Institute of Social, Economic and Government Research Report no. 31 (Fairbanks: University of Alaska, 1971)].

2. Arlon R. Tussing and Gregg K. Erickson, *Mining and Public Policy in Alaska* (Fairbanks: Institute of Social, Economic and Government Research, University of Alaska, 1969).

Comment

A. MILTON MOORE

I agree with most of Arlon Tussing's comments concerning the disadvantages of government ownership of mining companies. But I consider that his prescriptions concerning how British Columbia could gain greater control of its mineral industries could have been more specifically pointed to the British Columbia situation.

The first two circumstances he mentions as justifying government ownership of mining companies do not apply to British Columbia in the 1970's. Tussing's third reason is the offsetting of monopoly power by a private industry and the capture of economic rents. He might have added a fourth reason, namely, that sometimes government full or partial ownership of exploration companies may facilitate the achievement of or be necessary to achieve a socially desirable sequence and timing of the exploitation of mineral deposits.

British Columbia appears to be coping with all three problems fairly well with respect to natural gas; that is, coping with monopoly power, capturing the rents, and imposing some control upon exploration activity. The principal device used is the insertion of the British Columbia Petroleum Corporation as an intermediary between the producer and distributor. By setting the prices paid to producers and charging higher prices to distributors within the constraints imposed by political considerations and market conditions, the corporation seems to be well on the way to realizing the presumed objective of capturing as large a proportion of the rents as is consistent with inducing private companies to undertake the desired amount of exploration and development. The intervention of the federal government and certain provisions of the long-term export contracts have been obstacles to increasing the price of gas, but the corporation seems to stand a fair chance of overcoming them.

Control over prices and the ownership of most of the mineral rights should enable the government to adopt and implement a programme for the socially desirable timing and sequence of the development and production of its petroleum resources. In these circumstances, it seems to me that the main reason why British Columbia might want to set up a public corporation to explore for natural gas or develop new fields is that the incentives offered to private companies elsewhere, especially in the United States, might become so attractive that British Columbia could match them only by giving up an unattractively large proportion of the rents. That situation does not appear to have developed yet. Even so, it might be advantageous to have a crown corporation enter the industry's exploration and development stage in a small way to provide information, gain experience, and provide an alternative if private industry's price becomes unattractively high.

In addition, it would be politically advantageous to have a crown corporation in the exploration business. Exploration activity fluctuates for a variety of reasons, and it is often open to a company president or a newspaper editorial writer to ascribe a decrease in activity to burdensome taxation, exorbitant royalties or other charges for the Crown's mineral rights, or the government's hostility to private enterprise. A crown corporation could take up some of the slack when private exploration activity decreased.

Matters do not appear to be under the same control with respect to other mining. The provincial government appears to have been only partially successful in its endeavour to capture the lion's share of the windfall element of the economic rents. If the government had some crown corporations to turn to as an alternative, it would be in a stronger bargaining position to resist the attempt of private companies to retain the more favourable terms they used to enjoy.

Having argued that the use of crown corporations to conduct some of the exploration and development in all branches of mining would have some advantages, I agree with most of Tussing's comments concerning the problems of keeping public companies efficient and performing the function for which they were created. To Tussing's list I would add the caution that a government should decide before a crown corporation is created precisely how the company is to operate.

On the other hand, I am rather skeptical about Tussing's assertion that, where effective capital markets exist and there are a sufficient number of companies to create workable competition for resource rights, a government as landlord is more likely to maximize its revenue if it does *not* engage in production. His argument is that each tract put up for auction tends to be won by the bidder with the most favourable combination of expectations regarding recoverable reserves and development and lifting costs. This implies that most auctions are won by companies that overbid. If this is so, how is it that the industry continues to earn the supply price of capital? And why is it that the production stage of the petroleum industry returns much higher profit rates than refining or distribution? Surely, what matters is not the winning bidder's expectations as compared to the industry's average expectation but, rather, the winning bidder's expectations before exploration as compared to the outcome of the exploration.

Also, I do not know what the studies of the United States Outer Continental Shelf indicate. I do not know what conclusion to draw from the observation that successful bidders on the average earn a discounted cash flow on lease acquisition costs less than the oil industry's average rate of return on capital. I assume that the average rate of return referred to is on all other domestic operations, since it should not be surprising that the Outer Continental Shelf operations were less profitable than the operations of American companies in the Mid East. Even so, is it surprising that operations in a relatively new field are less profitable than in older properties? It is common knowledge that inventory gains due to inflation are a substantial proportion of profits during periods of rising prices. Finally, if the literature on industry behaviour under conditions of uncertainty is correct and the typical firm is a risk averter, bids for land rights should tend to be underbids and not overbids.

Tussing also mentions, in his list of important aspects of the problem of capturing the rents, the amount of geological information the landlord ought to obtain before opening land for lease or disposal. I understand that considerable uncertainty persists until much of the exploratory drilling is completed. If the typical firm is a risk averter, one might argue that the government as landlord should do its own subcontracting for the drilling of exploratory wells, instead of using mining companies as intermediaries.

Turning to the issue of the provision of capital, Tussing correctly stresses that sums in the hundreds of millions or even billions of dollars can be mobilized privately without government guarantees for projects in remote regions. I agree, but I should like him to remind private industry of this circumstance—as an antidote to the advertising to which we have recently been subjected. The argument pressed by the Canadian mining industry is that the exploration and development of the frontier resources, including those in northern British Columbia, will not be undertaken unless the operating companies receive the higher prices and profits that would generate the required capital internally. Since the general public is the landlord, the landlord is being asked to provide the industry with the capital they would use for exploration and development of the landlord's mineral properties. That strikes me as an odd way for the landlord to maximize his rents.

Let me close on a positive note. I particularly commend Tussing's final point at the end of his article. A government or public corporation should not attempt to do for itself the things that even the greatest oil and mining companies contract out to others, such as seismic surveying, core drilling, well drilling and construction—these are highly competitive subindustries.

Comment

LEONARD WAVERMAN

The article by Arlon Tussing on the role of public enterprise is well thought out as far as it goes. Unfortunately the paper does not go nearly far enough. Tussing's implicit view is that private enterprise acting alone would maximize social welfare in the resource industries. Therefore, he is very reluctant to see any usefulness for public enterprise, and where public enterprise is recommended, the conditions that Tussing suggests for its behaviour would have that public firm acting as a private one. This is where I disagree with Tussing. There is clearly a role for public enterprise in the resource industry, and that role is to behave very differently from private enterprise.

I am always surprised that public enterprise is viewed as an anomaly, a creature of strange origins, a mystifying inefficient beast, whose existence must be explained by some extraordinary irrational means. The fact that the government can offer goods and services whose output cannot be measured, (welfare, defence, conferences) as well as goods and services whose outputs can be measured (steel, telephone, coal products), I do not consider unusual. As we have grown accustomed to not having to justify government involvement in the defence industry, day by day, I am sure we will grow used to justifying government involvement in other industries. Tussing's suggestion that there are three reasons why government enterprise exists in a capitalist society—to maintain "unprofitable" activities, to amass vast capital requirements, and to collect monopoly profits or rents—strikes me as being unrealistic. The real question is how the political process allows certain groups or certain coalitions of groups to acquire the power of the state for their own purposes in individual countries. We cannot, in my view, define public enterprise as the provision by governments of goods and services which could be provided by a market economy and cut the list to include only things like steel or coal. What we are examining is how, in some countries, political power is acquired by the industry and used for its own purposes in setting up a regulated, private monopoly. In other countries, where unprofitable companies might fold and throw large numbers of employees on the public dole, the companies are taken over. In still other countries, instead of the companies being taken over, tariff barriers are erected; substitutes for the product are outlawed; strong pressures are placed to limit the immigration of labour.

There is then a very great difference between the reasons why public enterprise exists and why public enterprise ought to exist. If we are going to analyze a subject only in terms of its moral/intellectual level, we can argue about the cases where private enterprise operating in a competitive system will not lead to the social optimum; why the operations of government rules would be uneconomic and inefficient so that the intervention by the state directly in production is justified. Even if we debated on this level, a level I hasten to suggest is not the appropriate one, there are other factors I would add to Tussing's list and factors I would delete.

Tussing's first point is that externalities may require the continued operation of an enterprise which would make only losses when privately operated. This is the reason the state provides national defence, bridges, parks and highways. These are all goods which we may define as *public goods*, where the output is difficult to measure and where individuals can enjoy them collectively. Does the British government provide coal products because the externalities of reducing the industry size were immense? One must not only explain the existence of public enterprise in coal mining, but also explain its absence in other unprofitable industries. To do so one would have to step from our lofty moral level to the second realistic level and ask about the operations of the political process in Great Britain.

I think that Tussing's second point, concerning the imperfection of capital markets, is a red herring...(excuse the political comment). I really doubt whether COMSAT exists only because of capital market imperfections. I think that a number of firms would have jumped at the chance to be able to establish a communications satellite.

Tussing also states,"There is surely a bias against efficiency in most forms of state enterprise." Without further evidence on this point, I must disagree. Folklore indeed does suggest that private enterprise will be more efficient than publicly owned firms. If we are comparing the case of a perfectly competitive industry with the same industry owned and operated as a single firm by the government, then I would tend to agree with Tussing. However, when we are comparing public ownership with oligopolies or public utilities, then the issue is not clear. In fact, the political process may generate forces which tend to make public ownership more efficient than public regulation. Where a privately owned, publicly regulated firm wins large price increases, politicians are not directly involved. It is the firm's fault or due to the laxity of regulators. The public pressure exerted on regulated firms to minimize costs emanates from small understaffed consumer advocate groups. A publicly owned firm generates, as we have seen, substantial political pitfalls if corruption, excess payments, or rapid price increases occur. Public pressure through the election process may then force publicly owned firms to minimize costs so as to minimize trips to the public trough. Surely the question of relative efficiency should move into the more sophisticated framework of the way the internal reward and incentive structure of the firm affects decision making and risk taking. I would agree with Tussing that the public utility framework is most inappropriate for the oil and gas industries. To attempt to impose rate of return regulations on a diverse, and, perhaps, competitive industry would merely duplicate the enormous difficulty facing the Interstate Commerce Commission. Most experts are convinced that public utility regulation has failed. Currently, the minerals industry does not meet the criteria of a natural monopoly. A public utility approach would involve the worst combinations of both private and public worlds. If we are convinced that the industry is essentially competitive, and existing new forms of bidding or lease bonuses or royalty options will generate sufficient rents for the public, then the industry should be ignored. If we are, however, convinced that there are a number of social purposes which private enterprise cannot fulfil, then full public ownership is the answer.

This is where Tussing and I disagree. His view is that competitive industry

in the resource sector would, operating alone, maximize social welfare. I
believe there are a number of good reasons why private enterprise cannot be
expected to maximize social welfare. The two most important considerations relating to the role of a public enterprise are the questions of risk and
the optimal exploitation of resources. Tussing does bring up the case of risk
tangentially. He argues that the state would not maximize its return from
the ownership of mineral resources by mining them itself unless competitive
bidding were employed. But if the resource industry is risky, and if private
entrepreneurs are risk averse and therefore continually bid too low, society
would make greater profits by operating the mines itself. The Arrow-Lind
argument suggests that the government could then ignore risk since it is
pooling the ventures across all taxpayers. In fact, the government is not, as
Tussing suggests, operating one mine on one property; instead, it is
indulging in risk spreading and diversification by operating many projects
and many mines. The real question here is whether the level of government
we are analyzing is large enough for this diversification process. If we are
dealing with Canada, I would argue it would be; for British Columbia the
question is doubtful. The answer is elusive and must be analyzed not by
recourse to slogans but by recourse to empirical evidence.

If we really believe that mining is more risky than other activities, then that
is a major reason for government ownership of the industry. Therefore,
public enterprise could be the most effective means of maximizing rents.
Knowledge in any of the activities of the industry—exploration, development, drilling, refining, marketing—is not monopolized. Public enterprise
could hire the necessary factors of production. If private enterprise is
unwilling to develop resources because the private rate of return is too low
to compensate for the risk of bankruptcy, then let the state develop these
resources for there is no bankruptcy for the state.

In fact, there is much public enterprise in the resource sector. Disregarding uranium and electricity, consider the geophysical studies undertaken by
the federal government, the subsidization and training of engineers and
other labour forces for the resource industries, the sponsorship of research
and development by the government. It appears that private enterprise
wants the government to undertake the unprofitable infrastructure developments of the industry while leaving the profits for private enterprise. The
problem of information dissemination has already been discussed in other
articles in this volume. It is clear that there is a wide disagreement about the
optimal level of government dissemination of information, as well as about
the optimal level of government exploration. May I suggest that the case has
not been made that private enterprise operates most efficiently in conditions
of risk, high externalities in information production, and externalities in
production.

Even were we to ignore all the above factors which might tend to convince us that public enterprise is a useful option to consider, there are few economists who would argue that private industry operating in the resource sector provides the optimal rate of exploitation. Tussing suggests, "Canada is and expects to be a net exporter of mineral products." I do not know to which Canada Tussing refers. Private firms in concentrated industries have been large net exporters of mineral products. It is not obvious to me that Canadian social welfare will always be maximized by private enterprises exporting mineral products. As Robert Solow suggested in his presidential address at the American Economic Association meeting last year, the essential motivation for public ownership of extractive industries is the rate of exploitation. His case was made for a sector in an economy where there is no trade. I am convinced that adding a trade sector to his model would increase the need for government control of the rate of exploitation.

There certainly are means for the government to control exploitation other than by itself entering the industry producing resources. There clearly are means for the government to minimize risk in the industry and to disseminate information. Have the transaction costs of government involvement in regulation of supposed private industry and regulation of export and production permits reached the stage where it would be cheaper and more profitable for Canadian society if the government itself were to undertake development of these resources?

Even were we to agree that this stage has not yet been passed, the political forces already generated in this country make some government involvement in the resource sector inevitable. I would suggest that these government enterprises avoid reading Tussing's article. If they were to read the article (and follow the advice therein), they would only operate as private companies; for the reasons outlined in this comment, it is not private companies that we want.

Overlapping Federal and Provincial Claims on Mineral Revenues

JOHN HELLIWELL

How does divided jurisdiction affect the nature and efficiency of natural resource management? What are the effects of competition between jurisdictions? Is there any "right" division of authority and resource revenues between federal and provincial governments? These are some of the questions I shall address, but not fully answer, in this article.

There are four main sections in the article. In the first, I shall describe several major types of overlapping claims on natural resources, with emphasis on those that exist in Canada. In the second, I shall try to spell out how these overlaps affect the nature and operation of various types of mineral leasing policy. In the third, more aspects of the current Canadian situation will be presented as a prelude to the discussion in the fourth section of possible future developments. An additional concluding section briefly restates the evidence in partial answer to the initial questions.

The focus of the present paper is on provincially controlled mineral resources in Canada south of 60 °N. North of 60 °N is ignored because there is no established provincial jurisdiction over northern resources. I shall also pass over the special problems posed where provincial control is compromised by private freehold, the statutory limitations (in Manitoba, Alberta, and Saskatchewan) on the taxation of Hudson's Bay Company and Canadian Pacific Railway property, and similar limitations in British Columbia with respect to resources within the Railway Belt and the Peace River Block. These limitations, described in some detail by La Forest,[1] raise some complications for provincial resource taxation but do not alter the basic situation assumed in this article, namely that mineral resources fall under the control of the provinces.[2] Thus, in the sections to follow, the provinces are assumed to have the landlord's right to control natural resources, and the federal government's primary claims on resources are assumed to arise though their general powers of direct and indirect taxation and through the federal power to regulate trade and commerce.

GENERAL TYPES OF OVERLAP

Parallel Claims on the Same Tax Base

This is a classic form of overlap, usually resolved only by some form of revenue sharing agreement. [3] In Canada this situation arises primarily with respect to direct taxes; the corporation income tax being the chief jointly used direct tax applied to natural resources. With a firm revenue sharing arrangment, and with agreement about the details of defining the tax base, the tax overlap has no economic effects of any significance. Without agreement, the results can be chaotic, with total tax rates different from levels thought appropriate by either government and often likely to move in unpredictable ways. Even with agreement in principle, the two levels of government may have differing opinions about the desired rate of resource development and about the desirability of full collection of economic rents. This may lead to offsetting rate changes and other problems that arise without agreement about the appropriate combined rate of tax.

If the federal government is able to achieve agreement with each province separately, there are then problems raised by interprovincial differences in total tax rates. As described before, this can disturb the efficiency of resource use and rent collection if resources are spread over several provinces.

Parallel Resource Owning Jurisdictions

The best example of this situation is provided by interprovincial and international differences in the methods and levels of resource taxation. The overlap here is not in the tax base but in the markets for resources and resource capital. Suppose that one had worked out, in principle, the least-cost pattern of resource development to provide for a predetermined national demand. The resulting provincial pattern of economic rents would have to be collected to the same extent and by roughly equivalent methods for the least-cost pattern of resource development to be actually followed by firms free to invest in any province. If provinces do not share the same assessments about the desired timing of production, degree of development, and amount of economic rent available for collection, then their tax structures are likely to be inconsistent. If user costs are treated as zero by one or more jurisdictions, then competition among provinces for development projects is likely to lead to overly fast extraction and inadequate rent collection in all. If the extractive industries are fairly tightly organized and hard to enter, a province that overestimates user costs (relative to the estimates used by other resource owners) may find development capital abnormally scarce. This is only an oblique way of saying that if the factors used in the development of natural resources are themselves somewhat inelastic in supply (at least in the short run) yet mobile between provinces (or countries), then the price of these factors can be set high enough so as to capture for the developers some part of the pure economic rent, regardless of the technical efficiency of the royalty methods used by the provinces.

The extent to which this is possible naturally increases with the number of independent landlords and with their willingness to treat user costs as zero in setting fiscal terms. Because the popularity of governments is usually assumed to be improved by revenues or development now, the risk of user costs being treated as zero is substantial.

From the point of view of the mineral developer, the existence of independent jurisdictions competing for his capital increases the advantages of presenting a united industry front. The fact that the policies of provinces (and countries) are not really independent of one another, however, raises the spectre of large and unpredictable changes in fiscal terms taking place all over, adding to nature's uncertainties. It is conceivable, but not perceived as probable by the industry, that changes in fiscal claims by landlords should offset unpredicted changes in market prices and therefore act to stabilize net revenue streams accruing to developers.

Prior Tax Claim Affecting Residual Value

Any federal tax applicable to the production or sale of mineral resources affects the operation of any provincial leasing scheme.[4] The nature of the effects depends on the type of leasing system; detailed discussion is therefore deferred until the next section. The main federal taxes that will be discussed there are the corporation income tax and taxes on imports or exports.

Physical Control over the Movement or Sale of Resources

The landlord's right to develop may be limited or enhanced by restrictions set by another level of government. In Canada, energy resources provide the prime examples. The Federal National Oil Policy of 1961 put quantitative restrictions on crude oil imports and thereby increased the potential royalty revenues of the western producing provinces. By the same token, if ever the National Energy Board rejects an export permit that is approved by the producing province, the present values of resource revenues of that province as assessed by the provinces are lowered, unless the provincial export decision reflects a deliberate preference for development now at the expense of future revenues with a higher total present value.

Links between Resource Revenues and Various Transfer and Subsidy Schemes

Resource revenues potentially accruing to the landlord province may be either enhanced or diminished through interaction with federal subsidy and transfer schemes. The present value of resource deposits in an area designated for regional development grants is directly enhanced by the value of those subsidies. Resource values in full employment areas are indirectly lowered as a consequence, assuming some inelasticity of the supply of labour or capital used in development.

The key Canadian example, however, is provided by the complicated links between resource revenues and the federal system of equalization payments to provinces with less than average revenues. Under the Federal-Provincial Fiscal Arrangements Act of 1972, a standard per capita provincial yield is calculated for each of nineteen revenue sources by applying an average provincial rate to the actual tax base in the province. If, for any province, the sum of these per capita yields is less than the national average yield, then the per capita difference is paid by the federal government to that province. The recipient provinces have typically included all but Ontario, Alberta, and British Columbia.[5] One consequence of the present equalization system is that additional current resource revenues accruing to a have-not province lead to an almost equivalent offsetting reduction in the equalization payments from the federal government. On the other hand, extra resource revenues to a rich province like Alberta lead to extra payments from the federal government to the have-not provinces, unless the resource revenues are "deemed" to be capital payments to keep them out of the revenues used to define the size of the equalization payments.

EFFECTS OF OVERLAP ON SPECIFIC TYPES OF LEASING POLICY

Types of leasing policy will be considered under four broad headings: gross royalties, net royalties after deducting actual costs, royalties or payments based on estimated costs and revenues, and taxes or regulations governing resource use. These will be assumed to overlap with the federal corporation income tax. The effects of an export tax will be discussed in the next section.

Gross Royalty

The gross royalty is typically a proportion of the output of the mine or well, usually paid in cash on the basis of posted wellhead prices but sometimes taken in kind. On a flat rate basis, the gross royalty is not able to cope with resource deposits of differing quality and cost of extraction: any rate high enough for the rich deposits is too high for the poor ones. This problem also exists with the gas purchase methods currently used by the British Columbia Petroleum Corporation. A single price is used for all new gas; if it is low enough to leave only a normal return on the low-cost pools it is too low to cover the costs of developing needed marginal supplies.

The overlap between the corporation income tax and the gross royalty may be shown symbolically. Let R be actual gross revenue per unit of production, C actual per unit costs of development, excluding royalties and return to equity, while tf and tp are the proportional federal and provincial corporation tax rates, trg the provincial rate of gross royalty, and E the return to equity after corporation income tax. F and P are the total per unit returns to the federal and provincial governments, respectively.

If royalties are deductible under the corporation income tax, the returns to the three parties are as follows:

$$F = tf((1\text{-}trg)R\text{-}C),$$
$$P = tp((1\text{-}trg)R\text{-}C) + trg(R),$$
$$E = (1\text{-}tf\text{-}tp)((1\text{-}trg)R\text{-}C).$$

The sum of the three returns is simply R-C. If the provincial gross royalty rate is raised by 0.1 (10 percentage points), provincial revenues rise by 0.1 (1-tp)R, federal revenues drop by 0.1tf(R), and the equity return drops by 0.1 (1-tf-tp)R.

If royalties are made nondeductible under the corporation income tax, federal revenues rise by tf(trg)R, provincial revenues rise by tp(trg)R, and the return to equity drops by (tf + tp)(trg)R. If, prior to this change in the corporation income tax rules, the gross royalty rate was set high enough to collect all economic rents, then it would have to thereafter be reduced. The reduction required to restore the after tax equity return (E) in the face of a move to make royalties nondeductible would be (tf + tp) trg/(1-tf-tp), measured as an absolute change in the proportional rate of gross royalty. For example, if tf were 0.4, tp were 0.1, and the initial trg were 0.25, the gross royalty would have to be removed entirely to restore the after tax return on equity.

In the federal budget proposals of 6 May 1974, which were put in abeyance when Parliament was dissolved, royalties were made nondeductible, while the federal corporation income tax rate was lowered by 10 percentage points for oil and gas production profits and 15 percentage points for mining and mineral processing. There were also changes proposed in the application of earned depletion allowances and in the speed with which exploration and development expenses can be written off. The latter changes will be ignored in the simple example to follow, because they are of lesser importance and require a dynamic analysis if they are to be properly assessed. Using the oil and gas situation as an example, the combined effects of the two main May 1974 proposals on the per unit revenues of the three parties are as follows, where tf is the prebudget federal tax rate:

$$F = -0.1(R\text{-}C) + tf(trg)R,$$
$$P = + tp(trg)R,$$
$$E = + 0.1(R\text{-}C)\text{-}trg(tf + tp)R.$$

These expressions, which sum to zero, show that if the producing provinces use the federal definition of taxable income, their revenues definitely rise. The net change in the returns to the federal government and to shareholders depends on the levels of the tax and royalty rates and also on the levels of costs and revenues. If we use, as before, tf = 0.4, and tp = 0.1, the net changes become:

$$F = -0.1(R\text{-}C) + 0.4(trg)R,$$
$$P = + 0.1(trg)R,$$
$$E = + 0.1(R\text{-}C)\text{-}trg(0.5)R.$$

At a royalty rate of 25 per cent or above, federal and provincial revenues rise and the equity return falls. At royalty rates between 20 and 25 per cent provincial revenues rise while the federal and equity returns both drop. At average royalty rates below 20 per cent, federal revenues drop as long as R exceeds C. What happens to the equity return depends on the profit rate, because the royalty applies to the gross value and the corporation tax to profits net of all costs except the return to equity. In general, only high productivity wells are likely to benefit. For example, at a 10 per cent royalty rate, the equity return rises for all ventures where R exceeds 2C. If we note that Alberta royalty rates are on a sliding scale depending on well productivity, we can see that the net effect of the two major May 1974 budget proposals is to lower equity returns for all Alberta oil and gas producers, to raise provincial revenues, and to raise federal revenues if the average gross royalty rate exceeds 25 per cent.

In the case of natural gas production in British Columbia, the operation of the May 1974 budget proposals would be somewhat different, as there is no royalty levied and the province collects its share of gas revenues through the trading profits of the British Columbia Petroleum Corporation. The May 1974 budget proposed to tax these profits indirectly, by a method to be described below.

Net Royalty Based on Actual Revenues and Costs

With some qualifications, the mining profits taxes used until recently in most provinces were of this general nature. By allowing costs to be deducted before the tax or royalty is calculated, this method avoids the gross royalty's discrimination against lower grade mineral deposits. I shall distinguish four methods of collecting a royalty net of actual costs of development:

a. as a straight royalty, after deducting some or all costs of exploration, development, and processing;
b. as a tax on profits, defined net of all current and some or all capital costs of exploration, development, and processing;
c. by sale of minerals from private developers to a crown marketing agency on a cost-plus basis;
d. by direct development through crown corporations.

Depending on their specific provisions, these alternative methods can be made to have very similar or rather different effects. The first three share a prime difficulty. To the extent that allowance is made for all actual costs of development, the royalty base measures economic rent. The closer the royalty base comes to being a good definition of economic rent, the closer the rate must be to 100 per cent if economic rent is to be collected. But as the royalty rate approaches or hits 100 per cent, the private developers lose any incentive for efficiency. This does not mean that development will slow down; since all costs are allowed for, including the cost of capital, there may still be presumed to be a supply of development capital, especially if there are seen to be ample opportunities for undertaking expenditures that have a consumption component yet are deductible from the base for the 100 per cent royalty. The efficiency problems become worse if the discrimination against risky projects is removed by allowing tax or royalty credits for ventures in which costs exceed revenues.

To illustrate the effects of the May 1974 federal proposals to make royalties nondeductible and to reduce the rate of federal income tax, we shall use the same symbols as before. A new cost variable C_a is needed to represent costs allowed to be deducted from the base of the royalty or profits tax, and the rate of net royalty or profits tax will be represented by trn. Ca will be greater than C if the allowed costs include some return to equity, and it may be smaller if a number of costs deductible for the corporation income tax are nondeductible for the royalty calculation.

If the royalty or mining tax is deductible from the corporation income tax base, the returns to the three parties are:

$$F = tf(R-C-trn(R-C)),$$
$$P = tp(R-C-trn(R-Ca)) + trn(R-Ca),$$
$$E = (1-tf-tp)(R-C)-(1-tf-tp)trn(R-Ca).$$

In the simplest case, where allowed costs for the royalty are equal to those for the corporation income tax, the expressions reduce to:

$$F = tf(1-trn)(R-Ca),$$
$$P = (trn + tp(1-trn))(R-C),$$
$$E = (1-tf-tp)(1-trn)(R-C).$$

In this case, it can be seen that the net royalty is equivalent to a provincial corporation income tax with prior claim on the tax base. Making royalties nondeductible in effect eliminates this priority. Federal revenues are increased by tf(trn)(R-C), provincial revenues are increased by tp(trn)(R-C), and the equity return is reduced by the sum of these two changes.

Example calculations will use the federal budget proposals relating to mining, as the mining profits taxes in several provinces resemble the corporation income tax in their scope. For mining, the nondeductibility of royalties was to be offset by a 15 percentage point reduction in the federal rate of corporation income tax.

The net effects of these two proposals on the three returns are:

$$F = (+0.4trn-0.15)(R-C),$$
$$P = +0.1(trn)(R-C),$$
$$E = (+0.15-0.5trn)(R-C).$$

In this example, the direction of the effects does not depend at all on the profitability of the mine, because the royalty and the corporation income tax use the same base. Provincial revenues rise; federal revenues rise if the rate of provincial mining tax is greater than 37.5 per cent and fall if it is less; the equity return rises if the rate of net royalty is less than 30 per cent and falls if it is greater.

The above calculations relate to net royalties collected by methods (a) or (b) listed earlier. If they are collected, as in the case of British Columbia natural gas, by having the producing firms sell to a crown corporation at a price sufficient to cover C plus an allowed return on equity, then there is no explicit royalty rate. The May 1974 budget proposals were intended to bring the revenues of a provincial marketing corporation indirectly under the corporation income tax by raising the taxable income of the producing firms to include "fair market value" (p. 15 of the "Notices of Ways and Means Motions" accompanying the May 1974 budget) of the mineral resource sold to the crown corporation. The only existing federal claim on mineral revenues handled in this way is the reduced federal tax rate applied to the producer's equity return. This is presumably a lower tax return (by the 15 per cent abatement) than would be obtained if the capital had been invested in another industry.

If the resource revenues are collected by the use of a provincially owned crown corporation to develop and sell crown owned minerals, then there are no federal revenues at all, either with or without the proposals in the May 1974 budget. This result follows from the present practice whereby crown corporations do not pay income tax. We shall return to this issue in the fourth section of this article.

Payments Based on Estimated Costs and Revenues

Myriad schemes fall under this general heading. They differ from one another in three chief respects: (a) whether some or all cost and revenue components are estimated; (b) whether the estimation is done by the landlord or the lessee; and (c) the timing and form of the payments by the lessee. The chief advantage of most of these methods over those described in the previous section is that they are theoretically better able to collect economic rent while maintaining incentives for low cost development.

One of the potential disadvantages of systems operating with royalty payments based entirely on forecast costs and revenues is that discrepancies between forecast and actual results are received or paid by the lessee. If the lessor government is better prepared than the lessees to bear risk, there may be grounds for tailoring systems that try to separate controllable from noncontrollable outcomes and for basing lease payments on actual outcomes for the former and on forecasts for the latter. Some possibilities are illustrated by alternative treatments under (a) above. New symbols are needed—let C' and R' be costs and revenues as estimated by the project developers, and C'' and R'' be the independent estimates made by the landlord provinces. A straight lease auction, in which cash is paid for the right to develop and no subsequent royalty payments are made, provides an example where the lease payment is based on an estimated volume of discovery multiplied by the per unit profit margin $R' - C'$ estimated by the developer.[6]

If it is felt that costs are more controllable than revenues, then a system could be based on $R' - C'$. This leaves open the question of how differences between R and R' are treated. If a cash or royalty bid were used as a starting point, it would have to be based on separate specific estimates for R' and C'. During the lifetime of the agreement, discrepancies between C' and C would be the sole responsibility of the developer. Discrepancies between R and R' could be split on an agreed basis. If all the differences were absorbed by the landlord, then the Crown would be operating in effect a price support system. If the difference between R and R' were divided between the lessor and lessee, there would in effect be a proportional tax on unforeseen price increases and a subsidy on price decreases. The "superroyalty" in British Columbia's Bill 31 is a partial example of this, as it contains a 50 per cent tax on the excess of a current international price over 120 per cent of a specified base value (see the article "Mineral Leasing in a Private Enterprise System" by McPherson and Owens in this volume). The Bill 31 example is not a complete one because the superroyalty provision was not part of the leasing agreements, nor does the law contain any symmetric subsidy arrangements to cover situations when the price is below 120 per cent of the base value. As a consequence, the provision lowers the expected value as well as the variance of the R stream accruing to the mine developers.

The above examples of type (a) distinctions suggest many important issues relating to risk sharing, but they do not have clear implications for the overlap of federal and provincial jurisdictions. One final point before proceeding: the cost and revenue symbols used have referred to costs and revenues per unit, while many lease auctions and permits cover areas whose mineral content is itself very uncertain. Thus the uncertainties involved in using forecast cost and revenues to determine fixed payments for lease areas are even greater than if the leases are paid for on a per unit basis.

Let us turn to type (b) variations. Who makes the forecasts? If the developers, then bidding on either a cash or royalty basis is the usual mode of lease transfer. If the lessor's estimates R'' and C'' are used, then the government in effect sets an asking price based on estimates of the costs of an efficient operator and either fixed or market responsive revenue estimates. The parallel with stumpage estimates in forest management is clear, although for minerals the use of such procedures might be suitable only between the exploration and development phases.

The type (c) distinctions are the only ones that really influence the distribution of resource revenues between the federal and provincial governments. At one extreme, if the bidding is entirely in terms of tax payments that are not deductible from the corporation income tax base (mining taxes after 1976 without the May 1974 budget, and most current taxes and royalties with the budget in effect), then all of the resource rents are part of the corporation tax base, thus giving the federal government a primary claim. Current taxes and royalties also get into the base for calculating equalization payments, thus reducing payments to a resource producing poor province and increasing the general federal revenue claims on a resource producing rich province.

At the other extreme, if the system is entirely "front-end loaded," with an initial cash lease purchase payment and no subsequent resource taxes, then the payments are deductible from the corporation income tax base. Furthermore, the amounts received by the producing provinces may also be treated as capital receipts and thereby not affect calculation of equalization payments.

Between the two extremes lurk any number of intermediate alternatives. The impression I have and would like to impart to the reader at this point is that the provinces could arrange to receive lease payments in forms that are deductible from the corporation income tax base. In my examples, the extremes have also involved different timing of the cash payments from the lessee to the lessor; but I suspect that not much ingenuity would be required to break that link. We shall return to this issue in the fourth section of the article.

Taxes or Regulations Affecting Resource Use

In general, provincial regulations are designed to produce employment or conservation benefits within the province at the expense of lower potential revenues for the provincial and, usually, the federal governments. The federal regulations have the same goals at a national level, and they have the same general effects on potential revenues. The main types of measure are: processing allowances, subsidies, or regulations encouraging use of the resource within the producing region; trade restrictions; and trade taxes.

If a processing allowance is paid directly out of what otherwise would have been a tax deductible royalty, then it actually enhances federal revenues. On the other hand, if the processing activity is made a condition of the lease, then it lowers provincial or federal tax receipts, or both, depending on the variety of factors listed in earlier sections.

The impact of trade regulations or taxes depends on whether they are designed to discourage exports or imports. If imports are discouraged, as by the national oil policy, then revenues rise in the producing provinces and for the federal government, if it has a stake in the net resource revenues. Export taxes or restrictions have the reverse effects. Calculations of effects are not meaningful beyond particular cases. Some of the features and effects of the current Canadian trade taxes and restrictions will be described in the next section.

ASPECTS OF THE CURRENT CANADIAN SITUATION

Here I roughly summarize complex issues, some features of which have been described earlier in the article.

Crude Oil

The key elements are:

a. A two-price system, with the domestic price lower than the export price by a margin ($5.20 per barrel, for example, in September 1974) that is subject to negotiation. The terms of the original agreement suggest a gradual reduction in the difference between the two prices. Prior to the agreement in March 1974 which set a wellhead price of $6.50 per barrel of crude oil for one year, the wellhead price was frozen at $3.80 per barrel from September 1973.
b. A federal export tax introduced in September 1973 covering the difference between the two prices. It was initially split with the provinces but is now used primarily to cover the federal subsidy to importers of offshore crude oil.
c. Royalty rates that have been sharply increased in the main producing provinces and collected by means that are being contested in the courts on constitutional grounds. In general, a distinction has been drawn between "old oil" and "new oil," with much higher royalty rates on the former than the latter.
d. Export quantities set by the National Energy Board on a month-to-month basis, based on producers' surplus over demand forecasts by Canadian refiners. National Energy Board hearings in 1974 considered the future potential for oil exports; most submissions forecast an end to exports by 1980.
e. The May 1974 federal budget which, as described above, proposed several changes in the taxation of oil revenues, the chief ones being to make royalties and similar resource payments nondeductible and to reduce by 15 percentage points the federal tax rate on oil production profits. [7]

Most of these policies flowed from the sharp increases in world oil prices in 1973. The net effect of the whole situation has been to sharply increase the revenues of the producing provinces. However, there are indications that the subsidy from producing to nonproducing provinces implicit in the two-price system and the export tax will not be easy to renegotiate, especially when combined with the federal tax proposals of May 1974. On the other hand, for the producing provinces to get the current world price for their oil would create enormous pressures on the federal government and on the system of equalization payments.

Natural Gas

Prices and policies for natural gas are substantially different from those applying to crude oil, despite the fact that many present and potential users can switch quite easily from one to the other.

a. City gate prices for natural gas are still substantially below those for fuel oil in most parts of Canada.
b. Export prices for natural gas are less than half as high, on a heating value basis, as the oil export prices. This situation gave rise to National Energy Board hearings under section 11-A of the National Energy Board regulations.[8] The outcome of these hearings is not known but could be cabinet action to raise export prices, with or without an export tax large enough to keep natural gas priced comparably with crude oil in domestic and export markets.
c. In the absence of earlier action under (b) above, the British Columbia Petroleum Corporation was established to purchase gas directly from producers and to sell it at higher prices to domestic users, chiefly British Columbia Hydro, and hence to the export buyer required by contract to pay 105 per cent of the British Columbia Hydro price. These events are fully described elsewhere.[9] The key fact from the point of view of the present article is that the resulting profits of the British Columbia Petroleum Corporation have not been taxed under the corporation income tax. The intent of the May 1974 budget proposals was to tax these profits indirectly by imputing them to the producing firms.
d. Alberta gas prices have risen chiefly through redetermination clauses in existing contracts, and the province has obtained higher revenues chiefly from a sliding scale system of gross royalties.
e. The federal budget of May 1974 contained proposals for natural gas similar to those described above for crude oil.

Mining

a. Prior to the federal tax reform measures passed in Bill C-259 in 1971, the mining, oil, and gas industries all received preferential treatment, with mining the most preferred by dint of a tax-free period for new mines in addition to the percentage depletion and accelerated depreciation available for all the exractive industries. Since 1974, the tax-free period has been absent and the extractive industries have been treated similarly to each other, still with preference over other industries. The primary remaining advantages are the depletion allowance and the accelerated writeoff (immediate writeoff in the case of mining) of exploration and development expenditure.

b. Mining kept one significant peculiarity of treatment: according to an offer made to the provinces during the course of the tax reform debate in 1971, the provinces were given an abatement of 15 percentage points with respect to mining profits. In return, mining profits taxes were to be nondeductible expenses after 1976. Before that time, mining taxes of up to 15 per cent could be deducted from taxable income. Thus the federal budget of May 1974 contained mining provisions that were already slated to come into force in 1976. The only difference was that nondeductibility was also extended to royalties. This would not have been an important difference if the provinces had not started to use royalties for mining, as described below.

c. As a consequence of the looming nondeductibility of mining taxes, provinces turned increasingly to royalties which were still deductible as means of raising further revenues from the mining industry. In this chain of events, the May 1974 federal proposal to make royalties nondeductible can be interpreted as a foreseeable step in a federal-provincial chess game whose next moves are forecast in the next section.

d. All of the provinces with substantial mining industries have introduced or are studying substantial increases in mining taxation. [10] In part those changes were in response to the events described above, and in part they were in response to the sharp rise in world metal prices between 1970 and 1973.

e. The federal budget of May 1974 proposed a federal corporation income tax rate of 25 per cent for mining, made up of a federal base rate of 40 per cent abated by 15 percentage points.

POSSIBLE PATTERNS FOR THE FUTURE

First I shall consider the likely next steps if interjurisdictional conflict continues unabated, and then I shall consider some revenue sharing and equalization payment possibilities that might be of use if a more cooperative environment should evolve.

Continued Conflict

A possible pattern (with some implausible features) for a conflict ridden future is:

(F) Federal government reintroduces May 1974 budget unchanged.
(P) Provincial governments move towards the production and sale of all minerals by untaxed crown corporations, thus removing resource rents from the corporation income tax base.
(F) Federal government levies corporation income tax on provincially owned crown corporations.
(P) Corporations under federal government or Canadian Development Corporation control could be refused the opportunity to take part in the development of provincial resources.
(F) Federal government proceeds to develop the Yukon and Northwest Territories at breakneck speed, before they have time to become independent provinces.
(P) Provinces tighten restrictions on exports of energy sources and unprocessed minerals to other provinces. Resources consumed within the province are distributed at very low prices, with regulations to ensure that they be used for creating jobs and industry within the province.
(F) Federal government tightens controls on exports to the United States and other foreign countries, and takes over the mines and wells under section 92(10)(c) of the B.N.A. Act.
(P) Provinces shut off the mineral supplies at the wellheads and pitheads.

By this time, mineral production and development have ground to a standstill, and the courts are choked with constitutional issues arising out of nearly every one of the moves outlined above.

Searching for a "Rational" Stopping Point

If the confrontation mode described above is seen by the parties as wasteful as well as divisive, the chances for cooperative solutions may improve. If a conciliatory spirit does develop, any comprehensive and efficient solution ought to cover at least the following four issues:

a. Revenue sharing. The problem is to find a pattern that is consistent with the constitutional division of powers, efficient resource management, fiscal viability of both levels of government, clarity, certainty, and low costs of administration. From almost all these points of view, the preferred solution would be for the federal government to back right out of any special tax treatment for the resource industries. In doing so, the federal government would be operating in accordance with a sensible principle of constitutional interpretation that general powers (such as that of direct taxation) should not be applied unevenly to achieve specific objectives in fields (such as natural resources) that fall within the exclusive jurisdiction of the other level of government. [11]

Equal treatment of resource industries would require making royalties deductible expenses under the corporation income tax, removing the use of different federal corporation tax rates for different industries, and making capital cost allowances the same for all industries. The net effect of these changes on total federal and provincial revenues depends, as illustrated in the second section of this article, on the rates of royalty charged and on the profitability of mineral production.

b. Principles for operation of two-price systems. The present scope of two-price systems is rather broad. Crude oil is the central example, but two-price systems also embrace all commodities with domestic markets protected by tariffs or quotas or with export markets restricted by export taxes or regulations. Trade subsidies have similar effects. Agreed principles should be explicit about the reason for the system being adopted as well as contain a conditional timetable for its removal if it is intended to be temporary. A regionally specific assessment of the effects existing two-price systems would be a useful starting point for negotiations.

c. Agreed limitations on industrial use and subsidy. Resource rich areas may be inclined to treat locally abundant resources as free, even though they are scarce when seen from a larger viewpoint. To some extent, such regional differences are supported by transportation costs, and they are sometimes partly due to the existence of two-price systems created by export taxes or restrictions. For a mixture of reasons, provincial governments may provide resources to industries at even lower prices than justified by transport costs and agreed two-price systems. In part, this may reflect interprovincial competition for industry and, in part, an attempt to carry processing far enough within the province to get around export taxes and restrictions applying to exports of unprocessed and partially processed minerals. Examples of interprovincial competition include processing allowances in mining tax legislation that are far in excess of those required to make processing costs deductible from the tax base. Rules governing such allowances and other ways of using low cost minerals to purchase industrial growth must be subject to some broad forms of interprovincial and federal-provincial understanding if they are not to subvert and distort the good effects of agreements under (a) and (b).

d. Finally, the anomalous role of resource revenues in the equalization payments system should be altered. Ideally, all resource revenues should be included in the base used to calculate payments into equalization as well as entitlements for equalization payments. In another article[12] I briefly considered an alternative equalization payments system that involved a fund into which rich provinces make payments and from which poorer provinces draw. In setting the levels of entitlements, one of the aims would be to make the marginal tax rate (used for the equalization fund) on resource revenues equal for rich and poor provinces.

CONCLUSION

What can be said, after all this, about the questions posed at the beginning of this article? First, the division of powers in the B.N.A. Act has not in the past been adhered to closely in the case of mineral revenues. With the encouragement of the provincial governments, the federal government has used the tax system in attempting to increase the rate and alter the direction of development of mineral resources.

Second, federal-provincial competition over resource revenues was not a dominant problem when both levels of government were treating user costs as close to zero and were prepared to provide favoured treatment for the mineral industries. That situation has altered, and in the uncoordinated federal and provincial attempts to revalue natural resource deposits, the past history of subsidies is proving an impediment to agreed solutions. The provinces are making much over their exclusive constitutional rights to control resources, while the federal government is acting in the light of general powers and the precedents of using direct taxation to achieve specific allocation of mineral development and revenues. As shown in the second section of this article, the present links between provincial and federal tax and royalty systems are complicated and uneven.

Third, nothing is "right" when it comes to political division of authority and revenues, but I have suggested that it would be in the interest of efficient resource management and in the spirit of the constitution for the federal government to stick to industry-neutral direct taxation and for the provinces to act as collectors of economic rents from natural resources.

Notes

1. G.V. La Forest, *Natural Resources and Public Policy under the Canadian Constitution* (Toronto: University of Toronto Press, 1969), pp. 7-15.

2. Under the British North America Act, the provinces have exclusive lawmaking power with respect to "the management and sale of public lands belonging to the province and of the timber and wood thereon" (sec. 92.5), property and civil rights within the province (sec. 92.13), and local works and undertakings (sec. 92.10) excepting "works...Declared by the Parliament of Canada to be for the general advantage of Canada or for the advantage of two or more provinces" (sec. 92.10c). Abutting these specific provincial powers is the federal general power to "make laws for the Peace, Order, and good Government of Canada" (sec. 91) in matters not assigned exclusively to the provinces, and the exclusive power to regulate Trade and Commerce (sec. 91.2), and lands reserved for the Indians (sec. 91.24).

3. For a useful political analysis of Canadian tax sharing arrangements, see Richard Simeon, *Federal-Provincial Diplomacy: The Making of Recent Policy in Canada* (Toronto: University of Toronto Press, 1972); and for the arrangements themselves, see D.B. Perry, "Federal-Provincial Fiscal Relations: The Last Six Years and the Next Five," *Canadian Tax Journal* 20 (1972): 349-60.

4. The federal government is empowered by section 91.3 of the B.N.A. Act to raise money by any mode or system of taxation. The main provincial taxing power is under section 92.2, which empowers the province to levy direct taxation within the province. The main limitations on the taxing powers of both levels of government are found in section 121, which creates a free trade area within the Dominion, and section 125, which says that "No lands or property belonging to Canada or any Province shall be liable to taxation." Subsequent court decisions interpreting these taxation powers are analyzed in La Forest, *Natural Resources and Public Policy.*

5. For 1973-74 estimates of equalization payments and a brief description of the Fiscal Arrangements Act, see *The National Finances 1973-74*, published by the Canadian Tax Foundation.

6. The auction may be subject to reserve bids based on R''—C'', but this does not raise significant complications.

7. More precisely, the standard rate for oil and gas production profits was to be raised to 50 per cent, (including 40 per cent federal and 10 per cent provincial) from which a 10 percentage point reduction would be made. Thus the proposed oil and gas rate was 40 per cent compared to a post-1976 standard rate of 46 per cent for other industries outside manufacturing and processing.

8. Under section 11-A of the National Energy Board Part VI Regulations, amended September 1970, the board must report to the cabinet if there is "a significant increase in prices for competing gas supplies or for alternative energy sources" and the cabinet "may by order establish a new price below which gas exported [under licence]...may not be sold or delivered...."

9. A.R. Thompson and G.R. Armstrong, "The British Columbia Natural Gas Industry" (Paper delivered at the energy industry conference sponsored by the British Columbia Institute for Economic Policy Analysis, Vancouver, B.C., April 1974).

10. References to the various provincial bills are given in R.D. Brown, "The Fight over Resource Profits," *Canadian Tax Journal* 22 (1974): 315-37.

11. La Forest (in *Natural Resources and Public Policy*) quotes from Dunedin's 1932 Privy Council judgment in *In re Insurance Act of Canada*: "...if...legislation in form taxing is found, in aspects and for purposes exclusively within the Provincial sphere, to deal with matters committed to the Provinces, it cannot be upheld as valid."

12. J.F. Helliwell, "Extractive Resources in the World Economy," *International Journal* (of the Canadian Institute of International Affairs) 29 (1974).

Comment

ANTHONY D. SCOTT

The very interesting Helliwell paper speaks for itself and raises most of the questions that a concerned citizen would want answered before deciding where the power to tax should be assigned. I found the algebraic exercises particularly enlightening and difficult to improve on without detailed knowledge of the ratio R/C both in typical and marginal mineral properties throughout the country. Unless a firm's R/C position is such that it can benefit unusually from the promised drop in federal taxes, he concludes the expected outcome of the September 1974 taxes is that both governments will gain, shareholders will lose, and so new exploration will be somewhat discouraged.

But this outcome tells us very little about the best answer to the question, To whom should mining taxes be assigned? Recognizing this, Helliwell devotes the rest of his paper to aspects of that question. I might illustrate the problem by asking, Why should not the mining taxes and charges be levied by the provinces (or even by the municipalities) rather than by Ottawa?

One clue to the answer is given by Helliwell's distinction between gross and net royalties. We should recognize that in Canada both revenue bases are not available to all levels of jurisdiction. For example, municipalities may (generally) levy little more than a tax on property. This would not be a suitable or efficient tax for mineral enterprises, as the mining economics literature has long recognized. Again, provincial governments may not levy "indirect" taxes. It is true there are ways around this restriction, but it should be remembered that indirect mining taxes have been found to be *ultra vires* of the provinces, so the provinces have been compelled to have recourse to direct income or mining taxes on the one hand or proprietorial royalties on the other. The Canadian government, however, can tax any base it chooses. I conclude, therefore, that this freedom to use any tax base does create a sort of bias towards centralization in the Canadian constitution, to the extent that circumstances are likely to require forms of tax or subsidy that are constitutionally beyond the scope of the lower levels of government.

A second aspect of the mineral revenue assignment question is the impact of taxation on the timing of development and production. This matter is also dealt with by Helliwell. He writes,

> If provinces do not share the same assessments about the desired timing of production, degree of development, and amount of economic rent available for collection, then their tax structures are likely to be inconsistent. If user costs are treated as zero by one or more jurisdictions, then competition among provinces for development projects is likely to lead to overfast extraction and inadequate rent collection in all. If the extractive industries are fairly tightly organized and hard to enter, a province that overestimates user costs (relative to the estimates used by other resource owners) may find development capital abnormally scarce. This is only an oblique way of saying that if the factors used in development of natural resources are themselves somewhat inelastic in supply (at least in the short run) yet mobile between provinces (or countries), then the price of these factors can be set high enough so as to capture for the developers some part of the pure economic rent, regardless of the technical efficiency of the royalty methods used by the provinces.
>
> The extent to which this is possible naturally increases with the number of independent landlords and with their willingness to treat user costs as zero in setting fiscal terms. Because the popularity of governments is usually assumed to be improved by revenues or development now, the risk of user costs being treated as zero is substantial.

In brief, Helliwell asserts that the ability of firms to deny government a share of resource rent depends on the number of jurisdictions, their size, and their willingness to neglect user costs. This assertion is strengthened if he is correct that firms are more likely to collude against many small jurisdictions than they are against one or a few larger jurisdictions. I find this unconvincing. While agreeing that a given mining company is likely to attempt to play one small jurisdiction off against another (notice the word attempt), I feel that the incentives to cooperate are already so strong among mining companies in dealing with the government that the fact that "government" may be divided into many small jurisdictions is unlikely noticeably to increase them.

Helliwell also feels that small jurisdictions have short time horizons. This is of a piece with his opinion that small jurisdictions are particularly vulnerable to political interference from the mining industry. I am familiar with arguments of this sort and with examples that have been given of large mining companies completely dominating the decision making of relatively small jurisdictions. But such examples do not prove the point—far from it. The vulnerability of larger governments and of the central government to mineral or taxation proposals which appear to be politically expedient (which they find hard to resist because of their impact on employment or rural development) is surely as great as it is for smaller governments. Indeed, to go further, I might argue that when small governments bow before such pressure, they often know better what their own development needs and priorities are than do large governments (but this is a matter of opinion).

A third aspect of the assignment of mineral revenues is that of efficiency. In particular, we are concerned with "point of view." Helliwell refers to factor mobility between jurisdictions and, indeed, between countries. This leads us to ask, How important are efficiency conditions to Canada if alternative uses of labour and capital would be in other countries, other industries, or bring rents to persons who were not in the jurisdiction which is making the tax policy?

This is the classical problem of the policy maker in an open economy. What endowment of resources should he take as the proper constraint for his maximization of social income? On this question I am inclined to agree that small jurisdictions might well find themselves justified in promoting early or rapid exploitation. Their argument would be that there is no use making the best of labour or capital for exploration if the labour or capital would not be available to use in their own jurisdiction. Thus wasteful or excessive use of mobile factors locally may be better than zero use.

This possibility hardly tells against provincial jurisdiction. It's the same argument that restricts applying certain efficiency criteria to national policies. Ever since the time of the Carter Commission we have known that there was little point in arguing against excessive depletion or expensing allowances against Canadian income tax on the basis of inefficient use of capital, if the capital was not otherwise available in the Canadian economy. Thus I tend to discount much of the efficiency argument for placing resource taxation policies under the jurisdiction of the highest available government in Canada.

This brings me to the fourth aspect of the mineral taxation assignment question: distribution. The interregional redistribution question is elegantly and informatively dealt with by Helliwell under the heading of "Revenue sharing." In reading it I finally got some feel for the equalization formula now used in Canada. There are, in this field, basic and strongly held dogmas of territoriality and interdependence of consumption in Canadian political economy. The Helliwell position might be characterized as consistent with the feeling that output should be maximized for the country as a whole and then redistributed according to the preferences of the voters. (This principle corresponds to the separation of the allocation and distribution branches in the disciplinary field of public finance.) Applied to mineral taxation and the equalization formula, the principle reminds us of three problems. The first problem is a complete lack of consensus in Canada about how wide the area or region is over which output is to be maximized. I have already referred to this above as the "point-of-view" question. The second is, How wide is the social area of pooling for redistribution to be? A reading of the B.N.A. Act on this matter suggests that the fathers of confederation believed that the pooling area should be very small. Handing over "their" natural resources to the prairie provinces in 1931 also seemed to reflect a view that Canada's natural resources were to be used for the benefit of first comers or those who lived in the region where the resources were to be found. Most Canadians seemed not to disapprove of the idea that those who wished to obtain high incomes, or low taxes, should migrate to places like Alberta or Ontario, where natural resources can lead to high wage rates and relatively low taxes.

The third question is, What incomes are to be redistributed within Canada? Helliwell's discussion would suggest that rents alone are to be redistributed. On the other hand, the federal government's policy as embodied in the equalization formula evidently contemplates redistributing *all* incomes—not rent incomes in particular—toward some sort of national equality. I deduce this by observing that the basis for equalization payments is the income tax, and that the income tax collects these transfers on the basis of the height of individual taxpayer's incomes, not on the percentage of those incomes which comes from rent. The only effect of rent is a curious and indirect one: the greater the rents accruing to the treasuries of the wealthy provinces, the more equalization transfers national taxation must draw from wealthy income tax payers *throughout the country* to pay to the governments of the poor provinces. John Helliwell sees this complication as a problem, and recommends a scheme in which all resource revenues should be included in the tax base used to calculate provincial payments *into* an equalization fund as well as entitlement for equalization payments *from* this fund.

In considering this striking suggestion, I am worried by two possible defects. One is that resource revenues would not be included in the base unless each local province had decided to levy them. If, like Alberta in the past twenty years, it decided to leave many of the resource revenues with the mining and oil companies, then obviously such amounts will not affect the fund for equalization across the country. Secondly, as I have said above, it's perhaps too soon to argue that Canadians believe that resource revenues and rents should be distributed nationally. There seems to be some basis for belief that the expression "the rent of the land belongs to the people," made popular by Henry George, applies to the local people, not to the nation as a whole. A third difficulty is, of course, that the recipients of the redistributed funds proposed by Helliwell would be the taxpayers selected for relief by the recipient governments, not by the jurisdictions who were contributing to the redistributive fund. (This is an old British Columbia complaint.)

To summarize this potpourri of comments is difficult. The Helliwell article raises many questions in a useful and constructive way. My comments have been a short list of reflections touched off by the matters he raises.

In general, he concentrates on the question of which government should capture the rent. This is clearly put. But it appears to me that he simply assumes rather than argues that the efficiency criteria applicable to a Canadian jurisdiction should be the same as in a closed economy. Furthermore, I am worried that he assumes that the distributional goals for government are the capture of rents and their diffusion throughout the whole country.

Of course, I do not know or even argue that his assumptions are incorrect, but further discussion of the matters which he has raised requires an explicit axiomatic basis before economists can go much further in talking about other allocation or distribution in the assignment of tax bases.

Some Issues in Mineral Leasing and Taxation Policy: Preface to a Simulation Study

PAUL G. BRADLEY

When discussing alternative mineral resource policies, contemporary Canadian economists usually consider the government's primary goal to be the maximization of its own revenues. The analysis is invariably restricted to the mineral sector of the economy, and this goal is commonly described as maximizing the efficiency of rent collection. This terminology facilitates the assumption that progress toward the stated goal is consistent with income maximization for the entire economy. In this article revenue maximization is accepted as the appropriate primary goal to be served by a government's mineral taxation and leasing policy, and I examine some of the problems raised in defining and achieving this goal. The discussion is confined to the mineral industry, although it becomes clear that maximization of government revenues affects, possibly adversely, the level of national income.

It should be noted that economists recognize a variety of aims which can be served by resource policy. Only by wearing blinkers could this awareness be avoided, since in actual practice governments in Canada have always regarded other goals to be more important than the maximization of their own receipts from the mining sector. Traditionally, minerals have been bartered for economic growth. At present, accepted variants of this latter objective are regional development, either for its own sake or to achieve population decentralization, and the creation of a locally controlled industrial sector. However, either to suit the convenience of their models or to serve their grander schemes, economists prefer to establish the maximization of government receipts as the paramount objective of mineral policy.

The process of finding and mining deposits provides income gains for the economy in exchange for depleted natural resource capital. This is a continuing process, so a long-term view of revenue maximization is appropriate, in contrast with the notion of short-term rent capture. That is, alternative policies should be compared on the basis of the receipts they will yield the government not just from mines producing now but rather over a planning horizon which extends to include deposits yet to be developed or even discovered. When one attempts to judge alternative policies in this setting, large voids in the understanding of the economics of the industry become apparent. [1]

Reliable forecasts of the consequences of particular taxation or leasing policies require a thorough understanding of the economics of mining, embracing not only the circumstances of operating mines but also the prospects for new mines. Studies that would provide this understanding have not attracted much interest from government or academic economists. As a consequence, analysis of the effects of particular policies on particular industries is stymied.

This paper is concerned with one avenue by which analysis of alternative policies might proceed, the use of a model to simulate industry behaviour over a period of years. Some of the characteristics of the copper mining industry in British Columbia that are important and must be represented in a model are described. Principal issues that confront the province when formulating mineral leasing and taxation policy are also discussed. The final section details the basic components of a model currently being constructed and suggests what it may be possible to learn from the model.

THE RELEVANCE OF RENT THEORY

Economists approach mineral leasing policy with an instinctive urge to apply classical rent theory. Just as land was seen to be of differing quality for the production of agricultural products, ore bodies differ in their suitability for producing desired minerals. Just as it was once reasonable to see little other use for land than farming, there is now seldom an alternative productive use for a mine site in British Columbia. This conventional view is exemplified by Figure 1, which depicts observed supply costs for copper mines in the province, arrayed from lowest to highest.[2]

Figure 1 is derived from data describing ten copper mines currently producing in British Columbia.[3] Copper outputs are cumulated on the horizontal axis, while costs are shown vertically. The costs are average cost (per pound of copper) experienced over a one-year period. Operating costs of mining and concentrating the ore are shown on the bottom; to these are added capacity costs. The estimates of unit capacity cost require an assumption about productive life, here assumed to be twenty years. Figure 1 is based on data from diverse sources and must be regarded with some caution. Its purpose is to provide a rough notion as to whether variation exists in production costs among mines, variation attributable to the nature of various deposits which might give rise to economic rents in the classical sense.

Figure 1 portrays familiar, if oversimplified, notions about the existence of rent in the mining industry, where rent is defined as the return in excess of production costs. It suggests that output of an individual mine will not be reduced, even as virtually all rent is siphoned off by the resource owner.[4]

FIGURE 1

ESTIMATED PRODUCTION COSTS:
BRITISH COLUMBIA COPPER MINES

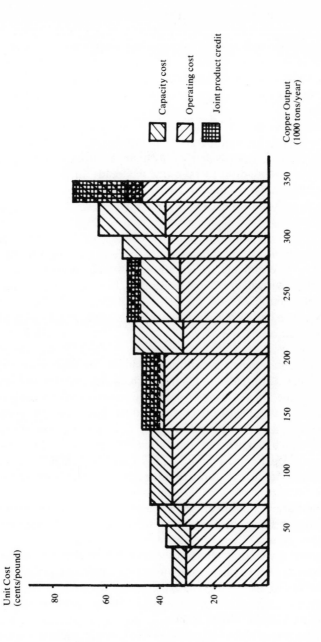

However, as industry supply is not inelastic, the total output may be affected by how a provincial landlord claims a share of the receipts. For example, royalties are often criticized as being "nonneutral," meaning that because they reduce the effective price which a mine receives for its output, some of the marginal mines (as shown in Figure 1) may cease to earn a return high enough to cover full economic costs. Where industry supply is upward sloping, a profits tax is often advocated because it will not change output.

Anyone familiar with the British Columbia mining industry can immediately suggest the limitations of the description of supply given in Figure 1. Its crucial deficiency is that it only depicts output where existing production plans, corresponding to investment already in place, are realized. If we consider supply in the longer term and allow plans to be revised, mineral output can be changed in several ways.

a. Deposits are not homogeneous. Greater amounts of ore can be obtained by exploiting lower grades not included in the original development plan; alternatively, lesser amounts of higher grade ore can be taken.
b. Other known ore deposits can be brought into production with the construction of new mines.
c. New ore bodies may be discovered, and these may also be brought into production.

The nonhomogeneity within individual deposits being mined in British Columbia is illustrated in Figure 2, which describes the ore contained in one zone of one mine. Grade of ore is indicated on the horizontal axis, and the cumulative volume of ore corresponding to a particular grade is shown on the vertical axis. Thus, while 45 per cent of the total volume of ore grades is better than 0.225 per cent, only about 12 per cent of the ore grades is better than 0.425 per cent. Figure 2 conveniently indicates the nature of reserves estimates, because it shows that they are necessarily dependent on what the mine manager believes the economic cut-off point to be.

What are the implications when the industry description of Figure 1 is modified because quantity of reserves varies with economic conditions, notably price? When price expectations change, individual mines, given time, will respond by changing their production plans. For example, when imposition of a royalty results in a lower effective price, the cut-off grade will be raised, as will the average grade of ore mined. Thus we might expect *high grading*: for a time, production of the desired mineral will increase, but the mine will shut down sooner. In the short run this is judged to be bad; capacity was in place and taxation has brought about a shutdown, although the cost of obtaining extra output was less than its market price. In the long run there could be circumstances where this high-grading effect would not

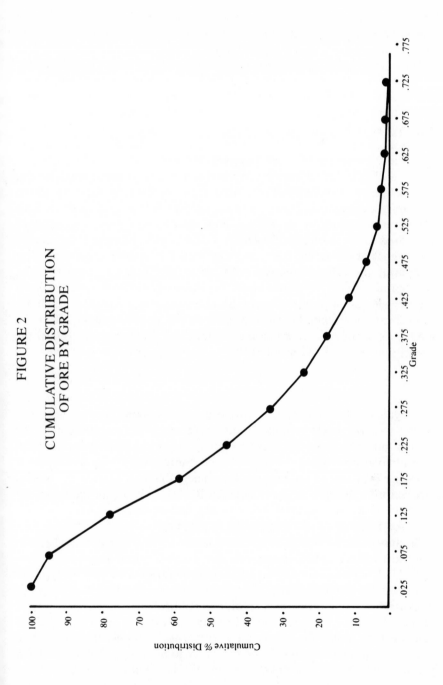

FIGURE 2

CUMULATIVE DISTRIBUTION
OF ORE BY GRADE

be disadvantageous. For example, in an industry producing for the export market when it is desirable or necessary to limit the growth of the industry, high grading might be a preferred form of output curtailment.[5]

In the long run, there are other sources of increased output which are not accounted for in Figure 1: additional known deposits can be brought into production and new ones can be discovered and subsequently developed. Given time, industry output is responsive to economic incentives at the extensive as well as the intensive margin. As the possibilities for changing supply become more complicated than those indicated in Figure 1, the concept of *capturing the rent* becomes obscure.

In the myopic view of Figure 1, price-cost margins, or rents, were defined with reference to current industry output, but in the long run they would have to be defined with reference to the supply of developed reserves. This would reflect opportunities in the mining industry of a particular province compared to opportunities elsewhere. To carry over into the long run the policy goal of capturing the rents, one would have to measure them by comparing the value of reserves with the supply; however, specifying the appropriate supply function would be extremely difficult. Moreover, it would seem that any mineral taxation and leasing policy—whether its short-run effects were neutral with respect to output or not—would affect the industry's supply of reserves and output in the long run. Thus the conceptual basis of a capture-the-rents policy comes into question.

RISK

A cynic might claim that a government's planning horizon does not extend past the next election. If this were so, future investment in the mining industry would not concern policy makers. This extreme position seems implausible, since the electorate probably takes a long-term view of job security if not of government income. If the level of current receipts affects future receipts, decision makers will wish to know the nature of the tradeoff. How do corporate development and exploration plans within a province respond to different levels of taxation?

Because returns to investment do not begin immediately a decision is reached to proceed with a project, and because, once they begin, they extend over a period of time, investment must involve risk. Various types of risk exist in mining. Three types can be recognized, which may be designated as commercial, geological, and political.[6] Commercial risk is not restricted to the mining industry. In terms of demand, it is not obvious which is more risky, the copper industry or, for example, the automobile industry. The United States' present capacity to build large cars which burn premium gasoline is not enviable. Then again, the copper content of the nodules waiting to be scooped off the ocean floor may cause anxiety for British Columbia miners. One measure of the level of commercial risk faced by the copper industry is the variation in the price of copper. Figure 3 shows that the copper industry has experienced wide price swings, especially in recent years.

FIGURE 3

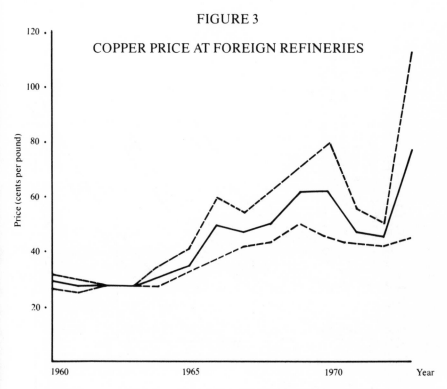

COPPER PRICE AT FOREIGN REFINERIES

Solid line shows annual average. Band between dashed lines shows range of quarterly highs and lows.

Source: *Engineering and Mining Journal,* "Annual Review," various years.

Geological risk does distinguish the mining industry from the manufacturing industry. At the exploration stage, drilling may reveal a very large, very rich ore body, or it may simply lead to the crossing off of certain areas on the map. Even after capacity has been installed at a particular site, the full potential usually only becomes known through a sequence of production plans, each preceded by drilling to determine the grade and structural characteristics of new parts of the deposit. Dimensions of uncertainty in exploration include the grade-size makeup of the deposit, its location, and the depth and structural characteristics which determine what type of

212

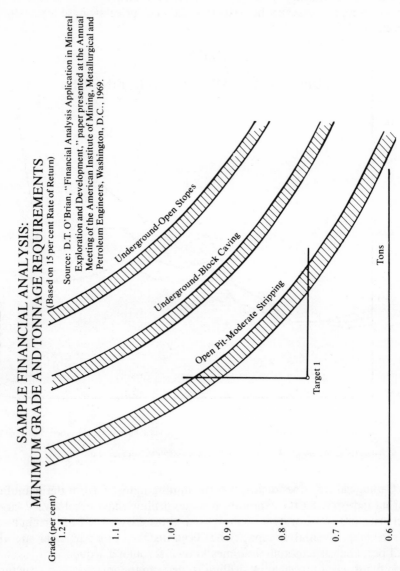

FIGURE 4

SAMPLE FINANCIAL ANALYSIS:
MINIMUM GRADE AND TONNAGE REQUIREMENTS
(Based on 15 per cent Rate of Return)

Source: D.T. O'Brian, "Financial Analysis Application in Mineral Exploration and Development," paper presented at the Annual Meeting of the American Institute of Mining, Metallurgical and Petroleum Engineers, Washington, D.C., 1969.

mining can be employed. If probability distributions can be defined for these characteristics, a combined probability distribution showing expected costs may be derived to characterize the risk in exploration.[7] Size and richness are both important and variable; since there are substantial economies of scale in mining, bigness can offset poor quality ore. Industry planning focuses on these two variables, as is shown in Figure 4, which defines a cut-off locus. Deposits which fall above this locus in both dimensions may be very profitable, but they are also very rare.

Geological risk in the form of variation in grade of ore provided bonanza possibilities for early prospectors, though success was elusive. For the modern mining company this risk is viewed as a problem in decision making under uncertainty. Risk is usually measured by the variance of the probability distribution of outcomes. Figure 5 shows the average grades of copper ore being mined in 1970 in Canada and in British Columbia. When one recalls that costs in the mining industry depend primarily on ore tonnages, the effect on earnings of having 3.0 per cent copper instead of 0.3 per cent copper is clear. In nature, of course, the frequency of occurrence of low-grade ore is much higher than indicated in Figure 5, since these data represent only commercial mines, and not all known deposits.

Political risk is often mentioned these days. When planning investments, companies must assess the probability that taxation and leasing rules will be changed. This category of risk is the least susceptible to quantitative measure.

A government acting with a long-run perspective ought to have, as observed earlier, information which reveals the tradeoff between additional receipts now from the mineral industry and lower receipts in the future because of reduced investment. The importance of risk in the industry adds another dimension that must be considered in framing provincial policy: How is risk divided under particular taxation and leasing policies? The panels in Figure 6 provide a simple illustration of how different methods of taxing the mining industry can change the burden of risk borne by the province.

The two upper panels in Figure 6 show fluctuating receipts over four time periods, attributable to variations in price. This represents what we have called commercial risk. In each period, output and costs remain constant. In the lefthand upper panel a 50 per cent profits tax has been imposed, and in the upper right hand panel, a 33⅓ per cent royalty is collected. The expected value of annual profits in each case is the same (mean = 2 units). The division of risk differs, however. Using the variance of receipts as a measure of risk, the company's profits have a lower risk under the profits tax than under the royalty (variance = 1 unit for profits tax, 1.33 units for royalty). The lower panels in Figure 6 tell a similar story for geological risk. Here four exploratory trials are assumed: the first two fail to yield a

214

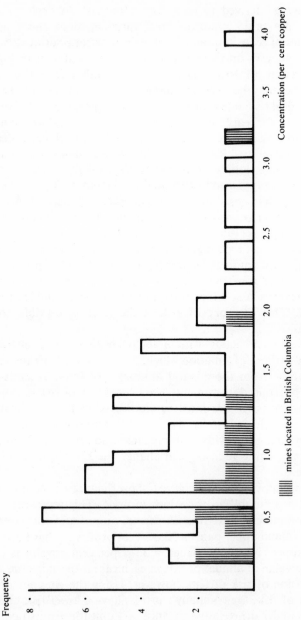

FIGURE 5

FREQUENCY DISTRIBUTION OF AVERAGE GRADE OF COPPER
AT PRODUCING CANADIAN MINES

▥ mines located in British Columbia

Source: *Canadian Minerals Yearbook,* 1970, p. 205-211.

FIGURE 6

VARIABILITY OF PROFITS AND TAXES
UNDER ALTERNATIVE POLICIES

Receipts Over Four Periods

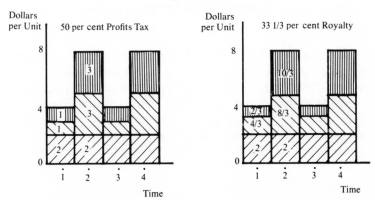

Returns to Four Exploratory Trials

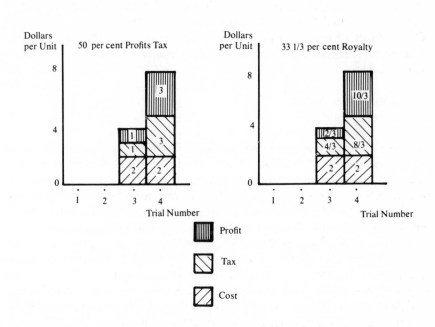

commercial mine, the third yields a mine capable of generating "low" before tax net earnings (2 units), and the fourth yields a mine capable of generating "high" before tax net earnings (6 units). We depict the same two tax policies as in the upper panel, a 50 per cent profits tax and a 33⅔ per cent royalty. The expected annual value of company profits is identical in both instances (mean = 1 unit), but the variance, measuring risk, for the company is lower with the profits tax than with the royalty (1.50 versus 1.89).

This simple example suggests that besides the tradeoff between present and future revenues from the mining industry, the provincial government must also consider the implications of any taxation and leasing policy for risk bearing. The more risk the provincial government is willing to accept, the less falls on the industry. Therefore, assuming risk averse corporate behaviour, policies under which the province shoulders more risk will stimulate corporate investment, leading to higher future provincial revenues.

The argument of this section is summarized in Figure 7. The vertical axis measures near-term government receipts from the mining industry. The horizontal axis measures long-term receipts. A capture-all-the-short-run-rent policy corresponds to a position very near the vertical axis: future receipts have been reduced to a very low value. In Figure 7 there are a family of curves depicting this present-future tradeoff. These curves correspond to the degree to which the provincial government accepts mining industry risk. Curves further from the origin represent greater risk borne by the province, and therefore higher expected receipts in the long term for any level of near-term receipts.

THE SIMULATION MODEL

After arguing that mineral taxation and leasing policy must be formulated with a long-run perspective, one has to admit the practical difficulties which this creates. A number of aspects of the British Columbia mining industry and of the mining industry in general have an important bearing on the results of any government policy. These aspects, which have not received much study, create some pertinent questions: What are the relative production costs of British Columbia mines, and hence the short-term economic rents? What is the elasticity of demand for copper and other British Columbia mineral products? What behavioural relation would permit us to forecast the responsiveness of investment in development and exploration to anticipated profits? How do alternative taxation and leasing policies alter the division of risk between the government and the private sector? And how might investment in exploration affect the order and

FIGURE 7

REVENUE POSSIBILITIES
FOR PROVINCE

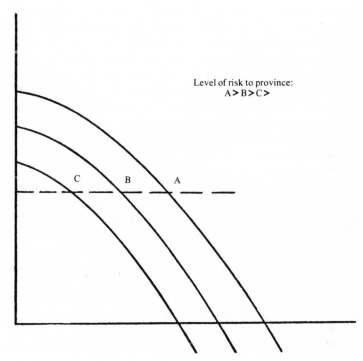

Long-term Receipts

manner in which deposits are produced, and what are the economic consequences? Answers to these questions would help to specify the tradeoffs depicted in Figure 7.

Research could rather quickly yield information on some of the questions in this list, but others are very difficult. In these circumstances simulation offers attractive possibilities. Where information about the British Columbia mining industry is not available, of course, simulation cannot create it. Simulation can, however, provide a flexible way of testing the consequences of various behavioural assumptions and of alternative taxation and leasing policies. Two applications of simulation techniques to the mining industry are particularly interesting: first, the physical parameters describing individual deposits can be specified so that output potential can be related to economic factors; second, risk can be taken into account by treating physical or economic magnitudes as random variables specified by particular probability functions.

A schematic diagram of a simulation programme is given in Figure 8. The structure is quite simple. In matrix E we record the physical and economic characteristics of known but unexploited ore bodies; in companion matrix D we record pertinent information about producing mines, such as remaining reserves, output, and costs and profits. The information contained in these matrices is updated period by period. Information for each mine is accumulated, so that at any time for each mine we can report cumulative output, government receipts (royalty payments, rentals, profits taxes), and corporate profits, and also the expected future values of these same magnitudes. At the same time, in each period, we accumulate aggregate industry data, such as total output, total government receipts, and total corporate profits. This accumulated industry information is recorded in matrix I.

The growth of the mining industry is governed by the rate of new capacity development and the rate of new ore body discoveries. In the diagram, the *development loop* box controls the investment in new mines. It is a feedback loop. For example, the amount of investment might be related to the difference between industry capacity in the preceding period and a trend line forecasting industry growth; alternatively, industry investment might depend on the difference between average British Columbia production cost and selling price. The *exploration loop* box similarly controls additions to the stock of discovered ore bodies, in other words, exploration investment. This is also a feedback loop, so that, for example, the number of new deposits found might depend upon the average grade of recent discoveries and industry profitability in recent periods. The feedback loop mechanisms which would best relate to actual industry experience can only be decided by further research; at the present we are compelled to rely on plausible assumptions.

A relevant example of the application of simulation techniques to mineral leasing problems can be found in the work of Azis and Zwartendyk at the federal Department of Energy, Mines and Resources.[8] These researchers developed equations relating costs and production rates to physical parameters. Then they examined the effects of changes in federal taxes on a sample of forty-two Canadian mines. Although we have followed similar procedures in developing equations which relate investment and operating

FIGURE 8

SCHEMATIC DIAGRAM OF SIMULATION PROCEDURE

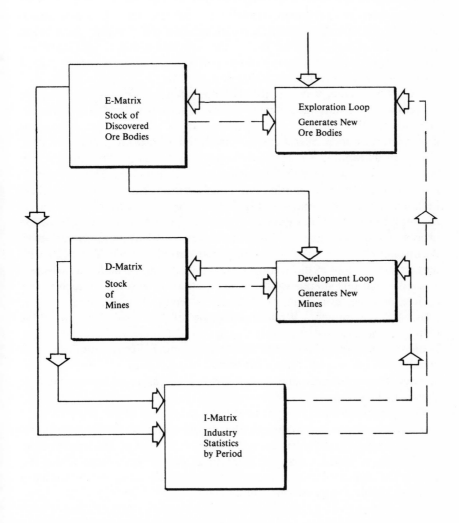

costs to physical parameters describing ore bodies, we are concentrating on simulation at the industry level and over time periods when new investment can occur. Over a twenty-year interval, with specified relations between profit incentives and industry investment in exploration and development, we may obtain comparisons of near-term and long-term government revenues from the mining industry under different taxation and leasing policies. With a dynamic industry model particular phenomena may appear in a different light. For example, the cost to the economy of high grading, unless demand and the supply of capital are highly elastic, depends upon the stock of alternative deposits available for development. If a new site is available, the cost (or benefit) of high grading will be contingent upon comparison of the grades of ore at the new and old sites and upon the magnitude of investment required to open a new pit, compared to that of instituting a new production phase in the old pit.

   We are particularly concerned with the need to take into account the uncertainty in planning taxation and leasing policy, and our model is designed for this purpose. Two previously described types of risk may be dealt with. Commercial risk can be introduced by regarding different price sequences as random variables. Industry development over a twenty-year period can be calculated for a particular price sequence. The calculation may be repeated for another sequence, chosen randomly, with these iterations continuing until a stable pattern emerges. Geological risk enters the long-run picture as exploration occurs. To simulate the exploration process, grade and size of deposit may be treated as random variables. Again, industry returns over a twenty-year period can be calculated, with numerous iterations as different values of the random variables are selected. These simulations should yield a stable pattern of results for each policy option, and the results may be ordered in terms of attractiveness to the provincial government. While comparison of the results of alternative policies is the principal objective of such a model, it is also intended for use in investigating the incidence of risk bearing under different policies. This enquiry will involve examining the variation, or "quality," of the streams of government revenues and private profits, as suggested by the example shown in Figure 6.

   The simulation model ought to show a great deal about the effects of alternative policies, even though the model will have to be kept fairly abstract to avoid making it so complicated that it becomes unworkable. Meanwhile, it appears that the state of economic knowledge about the British Columbia mining industry creates great difficulty in identifying a best policy for the provincial government.

Notes

1. This observation is from the vantage point of an economist, and it seems applicable to government tax authorities. The occasional technical papers which are published suggest that some of the sophisticated analysis carried on within the large mining companies would greatly help to fill these voids.

2. Figure 1 is not, strictly speaking, a long-run supply curve. The capacity, and corresponding unit cost, shown for each mine is that actually selected, with the cost of capital assumed to be 12½ per cent. A true long-run supply curve would show the *ex ante* range of capacity options. Depending on price anticipations, different levels of capacity with correspondingly differing unit costs would be selected.

3. The data on which Figure 1 is based were collected chiefly from trade publications and were discussed with industry personnel. This information will be presented in greater detail in a later paper. The mines depicted in Figure 1 accounted for 87 per cent of British Columbia's copper production in 1973. Copper is the principal product of each of these mines, although the mine depicted on the extreme right (the highest unit cost) is sometimes described as a molybdenum mine. For the three mines where byproduct values are significant (molybdenum in all cases), the shaded area indicates the reduction in the cost of producing copper which results if byproduct credits are assigned. Mines developed prior to 1966 are not included.

4. It is usually held that output will continue for as long as a producer is able to cover operating (or variable) costs. The situation in mining is not so simple, since the producer must also take depletion into account: there is a "user cost" that must be covered before production can be said to be profitable. The capacity cost figures shown in Figure 1 were imputed to satisfy the condition that if output were maintained at design capacity, all investment would be recovered together with an opportunity cost (discount rate) return within the planned life of the project. If the selling price just equals total production cost, and it is constant or increasing at less than the discount rate, the user cost will be less than the capacity cost shown; only if price is rising at greater than the discount rate will it be larger. In the absence of specified price expectations, it may be arbitrarily assumed that the producer will shut down if he is not fully recovering his investment.

5. It should be noted that high grading is a relative concept. Economic parameters will *always* determine a cut-off grade for a nonhomogeneous ore body, dictating use of only the higher grades. A form of high grading in which the richer grades of ore are used first is a typical mining practice in British Columbia porphyry copper deposits. During an initial production phase lasting perhaps three to five years, a higher grade of ore is mined and concentrated than the average present in the deposit. In the middle production years, ore from different areas of the deposit is blended so as to maintain a constant but lower percentage of copper in the concentrator feedstock. A further stage may entail providing a still lower grade of ore for the concentrator, perhaps including the tailings of the first phase. The production sequence, though constrained by physical availability, is determined in the light of economic calculations. This practice might be regarded as a form of "tilting" mineral output toward the present [see A.D. Scott, "The Theory of the Mine Under Conditions of Certainty," in *Extractive Resources and Taxation*, ed. Mason Gaffney (Madison: University of Wisconsin, 1967)].

6. This follows the classification scheme applied to investment in the petroleum industry in M.A. Adelman, *The World Petroleum Market* (Baltimore: Johns Hopkins University Press, 1972), pp. 53-56.

7. An example of such a procedure for the petroleum industry is given in P.G. Bradley and G.M. Kaufman, "Reward and Uncertainty in Exploration Programs," in *Arctic Geology*, ed. M.G. Pitcher (Tulsa, Okla.: American Association of Petroleum Geologists, 1973), pp. 119-25.

8. This work is reported in A. Azis and J. Zwartendyk, "Toward Optimizing the Public Gains from the Exploitation of Natural Resources," *Mineral Bulletin MR 129* (Ottawa: Department of Energy, Mines and Resources, 1972).

Comment

T.A. WEDGE

Bradley's article, although written at an early stage in the project, shows that he is heading in an important and potentially very productive new direction in the analysis of minerals taxation policy.

First, the project appears to be a solid attempt to narrow the communications gap and to integrate factual knowledge of the industry with economic analysis. The sources referred to in the article indicate that Bradley is circumventing the credibility gap by using mining industry literature which is intended for use within the industry, rather than as propaganda or negotiating material.

Second, Bradley has recognized that the theory of economic rent is not of great practical value in analyzing minerals taxation policy. The weaknesses in the economic rent approach seem obvious, but this approach has been so pervasive in other articles in this volume that a brief reexamination at this point is probably in order.

Traditional naive economic rent theory rests on the basic idea that there is a surplus return on resource production which originates from the quality of the resource. This surplus performs no economic function and therefore can be taxed away without lessening the supply of the resource product. This theory, however, requires the unrealistic assumption that technology is unchanging. But technology does change, creating resources capable of yielding a "surplus" return out of minerals which were once uneconomic. Developing this new technology has a cost, and the exploration which is required to find the new orebodies created by technology also has a cost. The original quality of the resource cannot be changed in any way, so, if any surplus remains after paying these costs, to argue that it originates from the quality of the resource is as pointless as it is to argue over how many angels can stand on the head of a pin. Since research on new technology and mineral exploration is probably undertaken in the hope or expectation of more than a normal return on the funds invested in such pursuits, it appears operationally impossible to determine whether any part of the supposed surplus can be separated from the incentives to conduct research and exploration.

This fundamental difficulty has been recognized by some advocates of the economic rent approach, who have attempted to solve the problem by including recognition of the costs necessary to stimulate the desired level of future production. It is then argued that existing institutions overmotivate research, exploration, and development expenditures, dissipating the surplus which could otherwise be obtained by taxation. It is asserted that this dissipation of the economic rent could be averted by withholding public lands from exploration or by increasing taxation so as to reduce the incentives for research and exploration.

This introduces a new problem: we are no longer talking about economic rent in the classical sense but using the term as an equivocation for what is really monopoly pricing. Instead of arguing that a nonfunctional surplus is automatically generated by the quality of the resource, we are really arguing that the government should create a taxable surplus by charging a monopoly price for the resource and keeping the monopoly profit, leaving the private producer sufficient income to produce the quantity demanded at the monopoly price.

At this point a very strong warning should be introduced. What reason is there to believe that the economic evils of monopoly pricing by private sector firms are not also characteristic of similar policies carried out by political jurisdictions? Politics is a complicating factor—greed may be as important a motivator to the taxpayer-voter as to the private monopolist.

If, in spite of this consideration, we do wish to take a monopolistic approach to minerals taxation, we should, like the business monopolist, be trying to maximize the expected present value of total profits. This is very different from the classical economic rent approach, which emphasizes something analogous to profit per unit sales, such as revenue yield per lease sale or per unit of mineral produced. Thus, the mistaken belief that we are still talking about economic rent can lead us to try to maximize the wrong variable—social return on sales—when we should instead be maximizing the expected present value of total social returns.

This necessary shift in the analytical emphasis should be obvious from the revised definition of economic rent, because the desired level of future production introduces a second variable, the size of the future tax base. This requires us to take account of the interdependency between the tax rate and the tax base, just as the would-be monopolist must consider the price elasticity of demand in his market.

Bradley has not only recognized the weakness in the economic rent approach, but he has made the shift in analytical emphasis. His approach includes the very important concept of elasticity of the tax base with respect to both the rate and form of taxation. The use of such a concept is essential if we are to reduce the currently significant economic risks of changing resource taxation policies.

Simulation modeling appears to be a valid methodology for applying the above analytical approach to actual policy analysis. There are always practical problems in simulation modeling, which Bradley obviously appreciates. Many readers will have very individual ideas about how to solve some of these problems or about what should go into such a model, but the beginnings of the model as described in the article form an excellent starting point for constructive criticism. In any event, it seems more important at this point to emphasize the value of his general analytical approach than to quibble about the fine details of the embryonic model.

One warning should be made regarding the use of the simulation modeling approach to the evaluation of mineral tax policy. When the form or level of taxation is changed, all the impacts on the provincial tax base may not be captured within a simple mining industry model. This is because different levels of mining output can also have an important impact on the taxable incomes of mine employees and of customer and supplier firms to the industry. The sales of transportation services, for example, are likened in a very direct fashion to the volume of physical output in the industry.

Apart from the great potential value of Bradley's approach in improving the quality of economic analysis of alternative minerals tax policies, such an approach may well also have great value in helping to close the communication and credibility gaps between industrial representatives and the economics profession. The process of obtaining the data necessary for building the model may provide opportunities for opening lines of communication between the model builder and the industry. And, intuitively, there seems to be a large probability that the tax policy which maximizes the social returns from mineral resource extraction will be one that the industry will be able to live with, because it will probably result in the long-run survival and growth of the British Columbia mining industry. This opportunity is really what the industry has been trying to preserve during the public debate over British Columbia's new tax policies.

PART THREE

British Columbia Leasing Policies

Mineral Leasing in a Private
Enterprise System *

J.L. MCPHERSON
O.E. OWENS

The theme of this volume is directed to mineral leasing, and yet the articles presented range over a broad topical area including public management of resources, public claims on mineral revenues, taxation, and the role of public enterprise.

Accordingly, this article is directed not only to the narrower theme of mineral leasing as it exists in British Columbia, but also to the point of view of the private sector respecting exploration, royalties and other forms of taxation, present British Columbia legislation, and public policy generally.

Particular reference is made to the metallic minerals field as contrasted to fuels and industrial minerals.

MINERAL LEASING POLICY IN BRITISH COLUMBIA

Situation Prior to 1973-74 Amendments

Prior to 1974, any person eighteen years of age and over or any registered corporate body could purchase a free miner's certificate which would entitle him to enter, locate, prospect, and mine upon crown lands for all minerals. The right to mine was secured by having possession of the free miner's certificate and title to a mineral claim of record, mineral lease, or, in the case of a freehold right, title to the crown grant mineral claim or other "freeehold right" to minerals. Of significance was the undisputed fact that the free miner had the right to mine the minerals so long as he retained title to the claim, lease, or "freehold," complied with the prescribed regulations, and paid the required fees and taxes.

Title was maintained by various combinations of payment of fees, annual rentals, annual acreage taxes, or the annual performance of a prescribed amount of exploration, development, or mining work. Fees were $25 per claim or lease unit; acreage taxes were 25 ¢ per acre; and work requirements varied up to $6 per acre.

*This article was written in 1974. Some of the observations it contains regarding effective and impending legislation therefore reflect the situation as it was perceived at that time.

No royalty was payable on minerals (other than iron and coal). A provincial mining tax was levied at the rate of 15 per cent of net income from all mining operations. To this was added the federal income tax.

1973-74 Amendments

During late 1973 and early 1974, changes were made to the legislation affecting exploration and mining in British Columbia. The exercise of the rights of a free miner was restricted to any Canadian resident eighteen years of age or over and to any Canadian corporation. The right to enter, locate, and prospect remains the same, and the right to mine remains on lands upon which the mineral rights had been alienated by grant or issuance of a lease. However, the right to conduct new mining operations became restricted, since it was made possible to engage in such operations only on leases issued or renewed at the discretion of the minister of mines and petroleum resources. Hence, mining was no longer clearly a right depending solely on the free miner complying with prescribed regulations and paying the required fees or rentals. Instead, this right was replaced by a permission granted by the minister, who could refuse to grant such permission unless, in his opinion, the operation, including its economic aspects, provided for the "best possible results from production." The minister's discretionary power under the act was not subject to any appeal in the courts.

Title to a mineral right includes the right to explore, but the right to mine became subject to the discretionary power vested in the minister. Title was still maintained by a combination of payment of fees, annual rentals, annual acreage taxes, or the performance of a prescribed amount of exploration or development work, but mining work was excluded. However, in the case of a lease, title could be suspended or cancelled at the discretion of the minister. Fees varied up to $50 per lease unit, rentals up to $2 per acre, acreage taxes up to $2 per acre, and work requirements up to $4 per acre.

Royalty rates proposed under the new Mineral Royalty Act on production of minerals rose sharply and were broadened in their application to include minerals designated from time to time by the minister. As royalties cannot be charged on production from mineral lands, that is, freehold rights, the new Mineral Land Tax Act provided for a tax on production from mineral lands to equate with the royalty rates on production from leases. The proposed royalty rate on minerals was to be 5 per cent of gross value, which was defined essentially as a net smelter return. In addition, a superroyalty was to be applied when the weighted average gross value exceeded 120 per cent of the basic value, and was 50 per cent of this excess. The basic value was initially determined by taking into consideration the average gross values during the preceding five years, and thereafter the basic value would increase or decrease annually to reflect any change in the value of the Canadian dollar or in the cost of labour, which, in the opinion of the cabinet, substantially affected the gross values of the designated minerals. If the basic value were to exceed gross value by 10 or 20 per cent, the royalty rate would be reduced by ½ and 1 per cent, respectively.

The provincial mining tax remained the same, so a cumulative effect arose from both the provincial mining tax and royalties and the proposed federal income tax. In the case of copper prices existing in 1974, some authorities [see *Canadian Tax Journal* (July-August 1974)] estimated the total tax burden as being in the order of 70-90 per cent of profits. In certain cases the tax take could exceed profits.

Major Points of Difference

The new or revised acts contained several points of major difference and of major importance to the mineral industry.

a. In Canada, mining law traditionally has granted anyone the rights to the minerals he discovered, so long as he fulfilled his obligations to society, paid his taxes, and respected the rights of his neighbour. Canadian farmers and Canadian homeowners hold their property by similar rights. The new mining legislation in British Columbia was based on the concept that mineral holdings are no longer the property of the owner but, in fact, belong to the Crown.

b. Security of the right to mine was lost. Admittedly, the free miner could acquire and retain ownership of mineral rights through staking a claim, performing the necessary annual assessment work, and paying the annual rental, but he had no assurance that after spending millions of dollars in exploration and development to prove a deposit, he could acquire the right to mine, since the minister gained discretionary power in the issuance and maintenance of a production lease.

c. The conditions under which a discovery may be put into production were not specified in law, as they had been previously. A decision became required from the political level of government regarding the terms of a production lease. The minister had discretionary powers and his consideration could include the economic viability of the deposit. This removed assurance that development of a deposit was possible after a discovery had been made.

d. High royalty rates were introduced. Their application from year to year could not be predicted. While the principle of royalty had been contained in the previous acts, it was applied cautiously and restrictively, in favour of a mining tax on profits or net income.

The minister or cabinet was thus given wide discretionary powers, and it became no longer possible for industry to assess the economic viability of a proposed or functioning mining operation on the basis of free market determination. A specific decision from the political level of government was required and such a decision could be subject to change from time to time, so financial projections became more uncertain than previously. As the Canadian Bar Association and many others pointed out, the language of the bill was not specific: items like *basic value* were not defined, and they believe that liability for a tax or an equivalent charge should be explicit and ascertainable and based upon a judicial interpretation of the will of the legislature, rather than an administrative discretion not subject to appeal.

We would like now to discuss certain general considerations we believe to be fundamental to the exploration for and development of mineral resources.

NATURE OF RESERVES

Mineral resources, that is, the concentrations of what are otherwise widely dispersed minerals within the earth, do not become resources in the economic sense unless and until capital, entrepreneurial skill, and labour are jointly put to work to create the value. Thus, the mineralization occurring in rock is simply an unknown geological situation unless it can be found and removed from the earth. An orebody is mineralization of sufficient concentration to permit economic extraction. Ore reserves in terms of the tonnage of rock containing a given percentage of the mineral are the means of measurement.

Ore reserves are not quantifiable on a permanent basis.

a. Their location or amount is unknown. They do not even "exist." They are created by exploration, changing market prices, technological development, and investments for development. The amount of ore present in any developed situation depends on price of product, operating costs, and taxes.

b. People tend to think of potential reserves as definite entities which will be available with some effort when required. This generalization is misleading for the reason that our future reserves must be found, and in a practical sense only so much can be found in any unit of time. Reserves do not lie in known situations waiting to be picked. If we take any extended period of time, say one hundred years, obviously within this time frame, in practical terms, the reserves that will be available are only those that can be found. The public wants to be assured of a dependable and adequate supply. Such a supply will depend on how much has been found and how much effort and skill is expended in the search for ore.

Since presently defined reserves are not sufficient for an extended period of time, we are dependent on finding more. It is not a straightforward matter to build up a successful exploration organization, whether by private industry or government. Experience shows that basic technology and skill is only acquired gradually. If activities cease or are substantially below requirements, the necessary organization and basis of information is not built up. It also takes time to build the mining industry required to extract minerals people will want.

c. Mineable reserves are not finite, because the development of a shallow ore body commonly leads to the finding of deeper or associated separate masses, and this discovery is only made by mining. The discovery of an orebody in one geological setting allows effective search in similar settings for others. For these reasons, it is a mistake to consider mineral reserves as finite in any time frame.

DIFFERENCES BETWEEN EXPLORATION FOR METALLIC MINERALS AND EXPLORATION FOR OIL AND GAS

The fundamental differences between exploration and development in the metallic mineral resources industry and in the oil and gas industry must be recognized in mineral leasing policy.

a. Most metallic mineral deposits occur in metamorphosed rock in a complicated structural setting. The deposits are minute in relation to their surrounding environment. In addition to being small, the deposits can lie at any angle relative to the surface. As a general rule, the greater the angle the smaller the target presented for discovery.
Oil and gas, however, occur in extensive, relatively undisturbed, nearly flat-lying, continuous sedimentary rock formation. The major dimensions are normally parallel to the surface, and the position relative to the surrounding rocks is predictable over relatively great distances. For these reasons, oil and gas present relatively large targets for surface exploration drilling.

b. The geological structures controlling the location of oil and gas are more predictable, which (1) simplifies discovery and determination of the extent of a producing field once found and (2) simplifies the location of other potential fields in the same type of geological setting. This makes discovery simpler, removes some risk, and even makes possible government auctions of adjoining ground.

c. Before extraction of a metallic deposit can begin, it is usually necessary to prove that surface drilling is representative and to determine bedrock conditions for mining. These are done by means of an expensive underground programme, which adds another large cost stage before economic assessment can be made.

ECONOMIC RENT AND ECONOMIC RESOURCES

The term "economic rent" has been lifted recently from the textbooks of economic theory and has been misapplied to the natural resource industries as the *raison d'être* for the method of applying an increased level of taxes and royalties. Economic theorists use the term "pure rent" as the return to land (in the traditional sense) which is fixed in supply with no alternative use. Hence, the value of land (the natural resource) and its return is determined by the market price, and taxes can be exacted without affecting the supply.

Economic rent is the term used by economic theorists to describe the return to any factor of production—usually labour or capital—which is temporarily fixed in supply. Hence, if the supply of labour is temporarily short, so goes the theory, in a particular occupation, wages will rise above the level which can be earned in an alternative occupation. This "surplus" is called economic rent, and, since taxes are an issue here, income taxes will rise accordingly. Such "surpluses" are temporary and usually disappear in the theory.

The misconception about economic rent applied to the mineral industry arises, first, because natural resources are not fixed in supply and, second, because natural resources do not become economic resources unless they are found and put to use. Thus, if economic rent, in fact, exists, it may well be in the profits arising from metallurgical and technological skills, refining and marketing assets, or entrepreneurial ability; or it may well be in wages as a result of superior skill, or bargaining power, or both. To conclude that economic rent exists and that it has been appropriated away from land is too facile a conclusion. Of course, none of this applies if there is no mineral to work upon.

RISK IN THE MINING INDUSTRY

The mining industry covers a broad spectrum of activities, from exploration, development, and the extraction of ore through the metallurgical process to fabrication, and sales. Each activity carries with it an intrinsic risk. The risk may be geological as it relates to exploration. It may be technical as it relates to developing the "find", for instance the technology appropriate to a particular grade or type of ore. It may be commercial as it relates to financing, production, and the probability of selling metal at a profit.

In general use, risk includes uncertainty; but in statistics the two terms are distinguished. Risk in the mining industry deals with probabilities that the industry deals with on a regular basis. Uncertainty is the introduction of the unknown. This is an important distinction in a volume on mineral leasing policy, dealing, as it should, more closely with the exploration aspect of the industry.

Industry experience shows that the average of its costs of finding a new mine is in the order of $25 million. If that mine is a large copper deposit in British Columbia, then over $100 million more is required to bring it into production. The financial risk (as well as the commercial risk) involved in bringing a new mine and metallurgical facility into production is very heavy. Not all mines are profit makers, at least for many years.

Contrary to popular opinion, mining does not have an unreasonably high rate of return. Mining companies are not exceptionally profitable companies. The return on capital employed by Canadian metal mining companies for the period 1969 to 1973 averaged about 9.5 per cent.[1] This return on mining was achieved under the previous more favourable federal tax system.

A mineral leasing policy is based on fundamental assumptions about the nature and occurrence of economic orebodies, the costs and benefits of discovery, and the public interest. In our society it is essential that, if we want something done, the rewards, considering the risk, remain sufficient to compensate for the work, the risk, and uncertainty as described. If we want it done efficiently, there is a considerable body of information to suggest that it should be done competitively with sufficient reward for the efficient. A firm belief of the existing mineral industry is that it can do this job.

Mineral leasing policy should not be developed in a narrow sense to apply only to the producing companies. It should encompass the prospector and the junior mining company, since their contribution and share in the ultimate gain is the determining factor in the level of overall provincial, regional, or national exploration for minerals.

INDUSTRY'S CONCERN WITH LEGISLATION

Certain changes in public policy towards the mining industry have been enumerated in this article as they affect leasing, exploration, production planning, royalties, and other taxes. We have highlighted the fundamental change in attitude to property rights which seemed implicit in the 1974 legislation.

A mining company requires security of title, security of tenure, clarity in "rules of the game" (that is, the terms and conditions under which it is to operate), with reasonable consistency, security, and regularity in tax treatment, and the prospect to earn a fair and reasonable return consistent with the risks of the operation. The company must see the opportunity to recover its overall exploration expenditures, the cost of developing and operating a mine, its fixed interest charges, and a reasonable profit. The situation created in British Columbia in 1974 did not meet these tests.

The 1974 legislation gave the minister wide discretionary powers. There was no assurance of consistency in the treatment of royalties and, hence, the overall level of taxation to be borne. The overall tax load of the mining industry should not be any higher than that of other Canadian industry, if the mining industry is to maintain its active role in our society. The language of the Act should be clear and precise, with uncertainties, particularly as to title to land and minerals, and to the rights of the free miner, elimated.

The attitude embodied in the 1974 legislation towards mineral leasing in British Columbia appeared to shift the determination of the value of a lease to the time when a decision is made to bring the property into production. This was a condition of utmost concern to the industry. Mineral leasing policy, as to share of returns, must be fixed before exploration is commenced. This is a fundamental point, otherwise it is hard to shake the entrepreneur from the conclusion that he is being unjustly denied what was offered to stimulate his activity.

Such conditions would make it extremely difficult or impossible to secure financing at any time, but they would make it particularly difficult in the future. The demand for capital funds in Canada and the world for the remainder of this decade is forecast at an extremely high level. Along with historical and prospective rates of return, this situation will shift the relative priorities of the investment community in placing new capital. It is useful to quote from the joint submission to the government of British Columbia by the Investment Dealers Association of Canada and the Vancouver Stock Exchange in 1974:

Summary of the Expected Effect if the New Mining Royalty Taxes are Implemented in Their Present Form

1. New mines will not be developed and existing mines will have reduced life.
2. Mining capital and skills will leave the province.
3. Investor confidence in British Columbia will be eroded.
4. Government plans for further development and job creation will be jeopardized.

The overall uncertainty that prevailed in the industry, and the delineated conditions that operated in respect to leasing resulted by September 1974 in a 60-75 per cent reduction in exploration activity in the province.

Canada has one of the world's most sophisticated mining industries in terms of such expertise as technological development, employment of skilled manpower, and entrepreneurial ability. The expertise in the exploration phase had begun in September 1974 to move out of British Columbia and out of the country. Consequently, the industry and the country was in the process of losing its ability to explore and to discover.

PUBLIC POLICY AND MINING DEVELOPMENT

Public policy has a major role in determining rates of economic development and in satisfying the public that they are receiving an adequate service from any industry. This service is achieved directly by supplying the product demanded by the consumer and indirectly through the financing of government programmes by revenues appropriated from industry by the public treasury.

Mineral leasing policy together with taxation influences what service the industry can provide by accelerating or discouraging exploration. The proper policy depends on a balance to achieve the broader economic development objectives of the country.

In determining a proper mineral leasing policy for Canada or for British Columbia, it is important to remember that the Canadian mining industry competes internationally. Our share of the world's known reserves is relatively small. Foreign reserves are often of a much higher grade and contain greater concentrations of metal than known Canadian ores. These geological conditions affect effort and risk evaluation. Exploration may be more attractive in areas of high grade ore or areas with good chances of finding new nearby deposits.

By and large, British Columbia is not an area with high grade ores, and does not have any known area to compare with the Zambian copper belt, for instance, which has a concentration of high grade deposits. There appear to be numerous deposits in British Columbia, but repeated discovery will be required to maintain output.

It is important that worldwide mineral leasing policy encourages the maintenance of output. Government policy sometimes works against this. The case of copper in recent years illustrates the point. Prices for copper have reached very high levels. Market demand has been high, but there have been interruptions in bringing new supply into the market caused by such action by governments as Zambia demanding control of equity in its major copper occurences; the Chilean government takeover of their very large copper production; and the Canadian government's changes in the taxation of the mining industry.

Each of these changes, which were initiated about 1969, upset plans for bringing major occurrences into production. This contributed substantially to a worldwide shortage of copper since 1973, and it contributed to the upward pressure on copper prices eventually paid by the consumer, in amounts probably substantially higher than the return to the taxing authorities.

Metal prices are determined in international markets. There is no way that they can be controlled locally or nationally, as the United States found out when it imposed price controls recently. When fundamental changes are made which affect the functioning of the market pricing system, the reaction may go far beyond the isolated feature considered in the change.

SUMMARY

The Canadian mining industry has an unparalleled record of achievement. It is viable and successful and has and is making a major contribution to the economy. To maintain this situation in British Columbia, mineral leasing policy must:

a. maintain in the law our traditional concept of property rights to provide security of tenure;
b. define in the legislation the terms and conditions of the leases;
c. apply these terms and conditions with consistency;
d. offer the prospect for a fair and reasonable return commensurate with risk;
e. provide consistency in tax treatment;
f. remove ministerial discretion as a prerequisite to the issuance of a mining lease and as a potential means of suspending it (except for noncompliance with terms of the lease);
g. provide the normal right of appeal to the courts.

The discovery of ore is not automatic. It requires the application of skills and money. The market place, while perhaps imperfect, does a very effective job of allocating relatively scarce exploration funds and skills.

Government is faced with choices in its dealing with development of mineral resources. Industry believes one of government's functions should be to encourage maximum recovery from each occurrence. Mining is always faced with the question of economic cut-off grades. Certain royalty policies increase these grades and turn ore to waste. Industry believes royalties should not be applied where they reduce recovery from mineral concentrations.

The government should give serious consideration to avoiding the multiple type of taxes on mineral development and consider obtaining its financial benefit through a single tax on earned income.

Notes

1. Return on capital employed is calculated by dividing net income after taxes plus interest, by the difference between total assets less current liabilities. All data were taken from Statistics Canada Catalogue 61-003 "Industrial Corporation Financial Statistics".

Comment

T.A. WEDGE

This article is a striking example of the wide communications gap which exists between the academic and government economists who are evaluating minerals policy and the representatives of the mining industry. The economists ask very pointed questions about existing policy but usually lack the detailed knowledge of the industry necessary to arrive at reasonably correct answers. The result all too often is policy conclusions based on incorrect generalization from other resource industries, such as forestry, fishing, or petroleum, all of which are different from mining in very significant ways. Another common error is the use of *a priori* hypotheses about the economic behaviour of the mining industry which are not adequately tested against the actual experience of the industry before being used to analyse potential policy changes. The mining representatives, on the other hand, have available much of the factual information which the economists lack, but they usually have not done enough homework on the economic theory behind the questions the economists ask to understand what information is relevant or how to present it in a form useful to the economists.

This communications gap has led to, and in turn has been complicated by, a credibility gap. Industry representatives are seen by economists, often justifiably, as self-interested special pleaders and complaining negotiators who contribute arguments and data which are self-serving and of little relevance to the questions being asked by economists. Economists, in turn, are seen by industry in a very derogatory light, because of the frequency with which their analysis contradicts the experience of the industry in the real world.

Both the communications and credibility gaps are seriously complicated by politics. Political issues have not been discussed much in this collection up to this point, but they have very important impacts on the economic behaviour of the mining industry which must be taken into account when designing minerals policy.

There is a serious danger that these gaps, also visible in other articles in this volume, may lead to major mistakes in economic analysis and in public policy decisions, which can result in declining economic performance in the jurisdictions where such decisions are made. Unless both economists and industry representatives make a serious and honest joint effort to close these gaps, important shifts in minerals policy will continue to entail serious economic risks.

Evidence of the communications breakdown in McPherson's and Owen's article is particularly strong in their discussion of economic rent theory. It is clear that the authors are not up to date with the economic literature on this issue. As a result, their argument only tangentially grazes the important weakness of economic rent theory as a basis for designing minerals taxation policies.

The communications gap is also reflected in weaknesses in organization and presentation elsewhere in the article. Nevertheless, there is much information in the article about the views of the industry which can be of great value to economists who wish to improve their analysis of the economic behaviour of the industry.

The article also contains very important clues to the impact of politics on the economic behaviour of the industry. The barrier which unpredictable uncertainty presents to private sector forecasting and financial planning is a very important one. Taxation is one of the more important variables in assessing the financial viability of mining projects, and it is very vulnerable to the introduction of unpredictable uncertainty as a result of political decisions.

From the industry point of view, the ideal is a system of taxation which is stable over long periods of time. In jurisdictions where such stability has existed and is expected to continue, the industry is free to evaluate projects on the basis of the probabilities it deals with on a regular basis. Any change in tax policy (even though it is to a level of taxation which should not unduly reduce the level of mining activity, and even though the new taxation approach is certain in its administration) introduces uncertainty from the viewpoint of the industry as to whether the new level or system of taxation will remain stable into the future. The more often tax policy is changed, the more the industry will fear unpredictable uncertainty of the tax variable. And as unpredictable uncertainty increases, the industry will require projects to yield greater returns to compensate for the risks. The result will be that fewer projects will meet the higher criteria, and potential mineral output will decline. The lower chance of exploration success due to the raised rate of return criteria will lessen incentives for mining exploration, which will result in further declines in potential mineral output.

The authors point out very clearly the concern of the British Columbia mining industry that the 1974 mining tax legislation created even more uncertainty than must inevitably result from any change in taxation, because the new tax system was no longer certain in its administration. The main source of uncertainty cited was the broad availability of ministerial discretion under the new legislation. What the article leaves unstated, however, is that this new area of uncertainty was severely complicated by politics.

The complication arose from the fact that the governing party in British Columbia at the time, the New Democratic Party (NDP), was generally understood to follow an ideology of democratic socialism. In 1969, a faction within the national party, believing that the party should follow a more socialistic platform, drew up its proposals in a document which had become known as the "Waffle Manifesto." This document stated that "capitalism must be replaced by socialism, by national planning of investment, and by public ownership of the means of production." The document's proposals for NDP policies respecting the resource industries began with "nationalization of resource industries," and the document later called for "immediate nationalization of resource industries and financial institutions...[as] the key to breaking corporate power in Canada." From the mining industry point of view, it was extremely significant that the NDP premier of British Columbia as well as some of the members of his cabinet had signed the Waffle Manifesto.

Also significant from the industry's viewpoint was a background and point paper on natural resources which was widely circulated in the mining industry. This document stated that "the acquisition of privately owned corporations in the resources fields prior to major resource tax changes would be a mistake because the market prices of those companies would be grossly inflated because of the wide range of tax holidays they presently enjoy." The document also suggested that an NDP government should "take over and operate firms unwilling and unable to function adequately" under a significantly higher tax structure, disclosure of all financial and other information the government deemed necessary, and worker representation on all decision making bodies. The document proposed public ownership of the resource industries and suggested the establishment of committees to do research on firms to develop feasible plans for public operation.

Finally, the *Cranbrook Courier*, a local newspaper, had published an interview with Leo Nimsick, NDP minister of mines, in November 1972. In this interview, Nimsick stated that "if private enterprise is serving the public in certain areas just as efficiently and just as well as it would be under public ownership, then we would apply our efforts towards the things that are not serving the public as well...I think the ultimate end is a socialist society, certainly...[but] conditions and circumstances will have to demand the changes.... We feel that we're not going into these situations just for the sake of public ownership or nationalizing something. It's going to be done for the benefit of the people in all cases."

The above documents were widely circulated in the industry. Taken together with the broad information requirements and the wide ministerial discretion in the new tax legislation, the industry had reasonable grounds to doubt that the true intention behind the legislation was a straightforward attempt to build a fair and reasonable structure of resource taxation. Many in the industry feared that the true intention behind the legislation was cheap nationalization of the industry following something like the following scenario.

a. Taxation on the mining industry is increased to a level which makes it impossible for private industry to operate efficiently, while the government claims publicly that the taxes are "fair."

b. The government claims that the resulting decline in the rate of mining exploration is due to industry collusion to blackmail the government, creating a need for a publicly owned exploration and development company to pick up the slack.

c. The required disclosure of information—and perhaps requirements for worker participation in management—allow the government to plan the takeover of existing firms.

d. Companies which postpone development of new mines due to the high tax rates on operations face high taxes on undeveloped deposits, and they are forced to surrender them to the Crown. The new mines are then developed under public ownership.

e. Taxes on mining operations are raised to the point where private industry is forced to shut down. The government then claims that nationalization is necessary to save the jobs of mine workers.

f. Nationalization is either at distress stock market prices caused by the inadequate earnings which have forced mine closures or at the salvage value of mining operations.

g. The industry is then run as one or more crown corporations not subject to taxation. This makes government management look efficient where private industry has failed, maintaining the political viability of the government and perhaps creating credibility for nationalization in other industrial sectors.

If this scenario, or something like it, represented the true intentions of the government, then the conference on which this volume is based would have been of little value for policy making in British Columbia—except, perhaps, to provide the government with propaganda material to "explain" or rationalize to the public the early moves in its cheap nationalization scenario.

If cheap nationalization was not the intention of the NDP government, it was still an entirely believable source of unpredictable uncertainty on the basis of the public record. Under such circumstances one would expect a lapse of several years before the industry were willing to plan new investment on the expectation of stable administration of the new tax system. During this lapse of time it would be unreasonable to expect the industry to make the normal major new commitments for expenditures on exploration or new mine development in British Columbia, even if the rate of taxation chosen preserved adequate incentives to undertake such activities. This time lapse led to another problem alluded to in the preceding article—you cannot put your expertise on ice, because it largely consists of expert people, whose abilities must be used if they are to be maintained and developed. This fact meant that companies marking time in British Columbia might have to use their expertise on projects in other political jurisdictions. The technical requirement and additional investment opportunities inherent in such projects might create an inertia problem in bringing the expertise back to British Columbia when the industry again viewed the political risks as reasonable.

Thus, if the NDP government honestly intended only to increase the tax yeild by raising taxes reasonably, it faced a major risk that the impact of political uncertainty would be stagnation and perhaps even shrinkage of the tax base, very likely to the point where the revenue yield would remain static or decline in spite of higher tax rates. If this risk were to be minimized, the British Columbia NDP government should have moved rapidly to forcefully dispel the cheap nationalization scenario, by publicly disavowing both the scenario and such documents and statements as those referred to above and by tightening and redrafting their mining legislation in ways similar to those suggested in the preceding article. Unless such action were taken, only minimal exploration could be expected by private industry in British Columbia—first, by companies completing exploration programmes to a convenient stopping point or to fulfil contractual commitments and, second, by the less risk averse entrepreneurs who would be willing to gamble on their expectation of the government being defeated in the next election. Only this latter group would dare to undertake a major new mine development project.

Apart from the very important political issue, the two authors of the article make a very worthwhile economic point—that the tax rate, whatever the form, can have an immense impact on the incentives of the industry to explore for and to develop the new mines which sustain and add to the tax base over time. As seen by the industry, the new NDP tax system in British Columbia resulted in a fundamental decrease in the expected returns from all mining operations. This reduced the number of known mineral deposits which were economically mineable in the short term and significantly reduced the economic incentive to undertake new exploration. Thus, even if the political uncertainty problem were solved, there appeared to be a major risk that the NDP government of British Columbia would lose more by shrinking the potential tax base than it would gain by increasing the effective tax rate.

Finally, this article repeats a very important point—that the royalty form of taxation chosen in British Columbia has problems from a resource conservation point of view, because it raises the variable costs of mining, thereby increasing the cut-off grade.

It seems clear in view of these issues—particularly the political problem—that one cannot easily dismiss the cutback in mineral exploration and new mine developments in British Columbia in 1974 as a collusive negotiating stance taken by the industry in order to obtain unreasonably lucrative taxation terms. It would be unrealistic to expect the industry as a whole to accept the evident risk of cheap nationalization, and it would be equally unrealistic to expect the industry to accept a lower rate of return on investments in exploration than it must earn to cover the cost of raising capital in the debt and equity markets. It is clear that there was a very serious risk of a major recession in mining exploration and development in British Columbia in 1974, if the political rhetoric and the form and rate of the new taxation were not seriously reconsidered.

Petroleum Leasing in British Columbia

DALE R. JORDAN

FACTORS AFFECTING THE LEVEL OF EXPLORATION

From data published by the provincial department of mines and petroleum resources, it would seem that exploratory drilling in British Columbia in 1974 was in a static position and may even have been declining. There was a 22 per cent drop in the number of exploratory wells drilled in 1973 compared with 1972, and a corresponding drop in the footage drilled. This decline in activity appeared to be continuing into 1974, with drilling down 10 per cent in the first seven months. This apparent decline in exploration for oil and gas in British Columbia came at a time when the demand for oil and gas was high. Prices for oil and gas had increased dramatically, and considerable concern was being expressed over national self-sufficiency in all forms of energy supplies. By comparison, exploration activity in Alberta showed the opposite pattern. In Alberta, exploratory drilling in 1973 increased 50 per cent over 1972. Statistics published by the Daily Oil Bulletin showed that during the first several months of 1974 there was a further increase over the same period in 1973.

What caused the static or possible decline of exploratory drilling in British Columbia at this particular period of strong demand and high price? There are several answers, and formulated theoretical solutions must be viewed in the light of political judgment where governments demand a greater share of resource revenues. Certainly, the federal budget proposals in the spring of 1974 discouraged exploration for oil and gas in Canada; however, it would seem that provincial government policies were largely responsible for any discouragement felt by the oil and gas explorer.

In this article I attempt to set out some of the basic causes that contributed to the 1974 situation in British Columbia. These causes are identified and any suggested solutions are offered in the realization that there is a danger of oversimplifying complex problems and the possible effects of implementing partial solutions.

For any analysis to be meaningful, a proper perspective must be maintained. One of the overriding factors influencing the oil and gas explorer's decision making is the number of geological prospects. In British Columbia, the potential hydrocarbon-bearing portion is thought to be restricted to the northeast part of the province. The western boundary of this 51,000 square miles is the Rocky Mountains. In 1974, an estimated 37 per cent of the oil and 18 per cent of the gas in this area had been found, leaving an estimated 0.82 billion barrels of oil and 47.56 trillion cubic feet of gas still to be discovered. Although there was considerable potential still remaining for the oil and gas explorer, the area was small compared to Alberta and Saskatchewan, and it was much smaller than the vast geological potential north of the 60th parallel.

Another factor that depressed the enthusiasm of the explorer was the apparent lack of multizone prospects. Drilling a well to test only one potential horizon increases the risk factor significantly.

Before leaving the subject of the province's hydrocarbon potential, it should be remembered that several basins existed, both offshore and in the interior, containing substantial deposits of sedimentary rock that have not been explored to any great extent. Generally, the known geology in these basins discouraged any extensive exploration. Possibly, the very expensive exploration needed in the basins required special consideration before their potential could be fully realized. This special consideration might take the form of a reduced royalty on any hydrocarbon discovered, or of a joint participation scheme whereby the British Columbia government might share in the risks involved.

Another factor to be taken into account by the oil and gas explorer when selecting his areas of interest is the access to those tracts that he feels have hydrocarbon potential. Again, British Columbia was at a disadvantage in 1974 when compared with Alberta and Saskatchewan. Access to this already restricted basin area was affected by muskeg conditions over much of the plains and by deep river valleys in the foothills. These conditions involve greater expense in conducting exploratory work, and the oil and gas explorer would often be confined to only a four-month work year, because of such terrestrial conditions. The problem of limited access presents a very real constraint on attracting the oil and gas explorer and is relevant when considering any changes in land tenure and revenue sharing.

Another important parameter in any deliberation affecting oil and gas exploration is the availability of funds for exploration.

As a general rule, funds generated from production are used to finance exploration. It seems that the risk factor employed by financial institutions effectively discourages the use of debt capital as a source of exploration funds. It is not uncommon to read that an oil and gas producer has arranged for a substantial loan, but this will nearly always be for a specific purpose related to the development of a new reservoir of oil or gas. The discovery of this reservoir would have been funded out of the company's cash flow.

When an oil or gas explorer's cash flow comes from production outside any particular province, then for money for exploration to flow into that province, that jurisidiction must have a framework of leasing and revenue sharing that can favourably compete with other areas also requiring exploration.

It would seem that cash flows from Canadian production will continue to be a major source of exploration funds. In the case of the integrated companies (those having refinery capacity), this will be true under almost any condition short of expropriation.

However, within this framework these funds will generally gravitate to where they can expect the greatest return. It may not be enough for a province to show that it will ensure that the successful explorer will receive a reasonable rate of return. It may well be that the successful criterion for increasing the level of exploratory drilling will be a division of revenues such that the oil and gas explorer will have a sufficient cash flow for an active exploration programme, as compared with other jurisidictions. The explorers who have a cash flow generated from production and the flexibility to determine the best place to reinvest these funds have the responsibility of ensuring that they can go to where they can expect to return the greatest profit.

The ability of the producer to channel cash flow from one jurisdiction to another depends largely upon the particular laws in force where the oil or gas is being produced. The provinces of Alberta and Saskatchewan tend to attempt to discourage the outward flow of funds generated from production in their particular provinces. In both provinces, this discouragement took the form of a high royalty, coupled with a drilling incentive programme, designed to encourage continued exploration.

In Saskatchewan, the mineral tax and the royalty surcharge took away from the producer all of the recent price increases and, in fact, returned to the producer a smaller amount per barrel than he was receiving prior to the price increases. As an inducement to continued exploration in Saskatchewan, an incentive programme was developed, whereby the producer was allowed to retain an additional 30¢ for every barrel produced, providing this money was used for drilling wells, for waterflood projects, research, and other specified purposes in Saskatchewan. This incentive had not been in effect long enough in 1974 to permit a complete analysis of its performance; however, it was really restricted to only those companies that already had production in Saskatchewan. A review of these companies shows that the producers with the majority of the production were the so-called Majors, which raised the question as to whether or not these Majors were prepared to continue to explore in Saskatchewan under any conditions. It is more probable that the exploration philosophy of the

Majors directed that they use their exploration funds searching for high reserve reservoirs (which generally means exploring outside Saskatchewan).

To encourage explorers who do not have production in Saskatchewan, the government provided an incentive credit of about 30 per cent of the costs of drilling exploration wells. This credit could then be used to reduce any royalty or mineral tax obligations that may accrue.

In Alberta, in 1974, there was also an incentive programme for exploration drilling. This incentive took the form of a credit which could be subsequently used to satisfy most of the cash obligations that may arise by virtue of the royalty obligation, rental payments and mineral taxes, and could also be used to purchase oil and gas leases. Credits established in Saskatchewan could not be used to purchase leases. The formula which determined the amount of credit that could be established for any particular exploration well was predicated upon the area of the province in which the well was located and the depth of the well. This formula was expected to return to the explorer by way of credit approximately one-third of his drilling costs.

The principal producers of crude oil in Alberta were much the same as in Saskatchewan—they were the Majors in the oil and gas industry. Again, as in Saskatchewan, it would seem that the interest of the Majors in continuing to explore in Alberta had become blunted, not because of any particular rules, but more through the apparent lack of sufficient high reserve potential which this type of company must search for.

In Alberta, in 1974 the royalty was structured to increase not only when the production increased but also when the price increased. Alberta had also adopted a new oil/old oil concept, whereby royalties were considerably reduced on what was termed new oil, which, of course, had the affect of increasing the producer's cash flow and encouraging the development expenditures. This old/new concept applied as well in the case of natural gas. Saskatchewan also had provision for a reduction in its mineral tax and royalty surcharge for new oil. This reduction was gradually phased out after a few years. Alberta, in 1974, had the lowest rate of royalty on oil and gas, received the highest price for its natural gas and had a price for crude oil equivalent to the other provinces, all of which means that in Alberta the producer of oil and gas received a higher rate of return and, consequently, had more money available to him for exploration purposes than the producer in Saskatchewan or British Columbia.

While the incentive programme adopted in Alberta and Saskatchewan offered encouragement to drill exploration wells, any analysis of the performance of a similar programme which might be suggested for the province of British Columbia must be coupled with the consideration of a lower royalty and subsequent higher cash flows to the producer.

This type of incentive can only be effective if the producer is also offered a return on his development expenditures that is competitive with other jurisdictions into which the oil and gas explorer is free to go.

Another source of money for drilling exploratory wells previously used in North America has been the drilling fund. The drilling fund usually takes the form of buying a number of shares in a limited partnership. The attraction to the investor apart from the possibility of participating in oil and gas discoveries, is the income tax feature in the United States which permits all intangible drilling expenses to be written off in the year that expenditures were made. This can be done without qualifying under any principal business rules, as prevails in Canada. It is this principal business rule that has to the present time precluded the tapping of a similar source of exploration funds in Canada.

It is estimated that between 250 and 400 million dollars are generated annually through the sale of shares in drilling funds in the United States. The United States government, in an attempt to encourage further exploration within its own borders, has chosen to reduce the amount that can be written off against income for income tax purposes when the funds are spent outside the United States. This move precipitated the drying up of exploration funds in Canada which previously came from this source.

Over the past few years other extractive industries, particularly mining, have been channelling considerable amounts of their cash flows into oil and gas exploration. These endeavours have usually taken the form of funding a subsidiary company. The continuation of this type of fund for the purposes of exploring for oil and gas will depend upon the success of the subsidiary companies and also upon the influence that mining taxes will have upon the parent companies' cash flows.

A few years ago it was quite popular for large United States gas utility companies to provide exploration funds on the basis that they would have first call on any gas produced and exported for sale to the United States. This source of exploration funds has also dried up almost completely as a result of decisions by the National Energy Board concerning the exporting of gas and, as a result of rulings by the Federal Power Commission in the United States, which have not allowed these utility companies to include these expenditures in their rate base. The Federal Power Commission looks upon this type of exploratory funding as a mortgage loan, rather than as a prepayment for gas, as the utility companies would prefer.

THE LEASING SYSTEM

The 1974 system of leasing in British Columbia will now be reviewed, and I will attempt to analyse its effectiveness in relation to the relatively small hydrocarbon potential area, lack of access, difficult terrain, and exploration funds generally coming from cash flows.

The system of granting oil and gas rights on crown lands in British Columbia invoved in 1974 an exploration grant called a "permit," a subsidiary exploration grant called a "drilling reservation," and a development grant called a "lease."

Permits, which can involve upwards of 100,000 acres, could only be acquired through a competitive cash bidding system at sales which were usually held four times a year.

The permits were classified from A to D, depending upon their accessibility and the terrain conditions. The class D permits were for offshore areas. One purpose for classifying permits was to allow them, where the working conditions were difficult, to have a longer life than was otherwise provided.

Permit classification also determined the minimum amount of work the permit owner was obligated to spend in any year. These minimum work obligations were more stringent for the Class A permit, where access and terrain problems were minimal. The significance of this particular requirement was lost because, in order to conduct an equivalent amount of exploration, particularly drilling, the costs involved in the offshore areas far exceed those which would be expended on a class A type permit. The sections of the Petroleum and Natural Gas Act that governed the permit work obligations did not specify any specific exploration programme, but only that a certain amount of money be spent during each term of the permit. This obligation to spend money in exploring on a permit could be satisfied by grouping several permits together, so that expenditures incurred in exploring on any one permit would satisfy the work obligation of the grouped permits. This grouping provision is important because it allowed an explorer to acquire large tracts of land for a short period in order to conduct extensive geophysical exploration.

This obligation to spend money on exploring on a permit could also be satisfied by paying the money to the British Columbia Crown. This would seem to be about as negative a provision as one could imagine, assuming, of course, that the government of British Columbia was interested in ensuring that the companies holding permits were the ones prepared to actively explore. If the removal of this system would give rise to problems, as in the case of an explorer unable to work a permit through no fault of his own, then a far better system would be to provide a means where work obligations could be accumulated and satisfied in the following year.

Permits were valid for one year and could be renewed annually, for a period ranging from five to eight years, providing the company was not in default.

When a permit holder had expended the minimum amount of money exploring on a permit or group of permits or had paid the money to the government, he was entitled to convert the permit into leases. To do this the permit holder had to relinquish 50 per cent of the land back to the Crown. Land selected for conversion into leases could not be in a consolidated block, but had to be in a number of leases, which had to corner one another, or be separated by at least two units—approximately one mile. The maximum size of the lease was six units—approximately three miles square.

This system of exploratory permit and subsequent conversion of half of the land to lease on a chequerboard fashion was similar to that used in Saskatchewan and part of Alberta. The system probably had its beginning in Alberta. It was designed to ensure that when crude oil was discovered, the Crown would be returned some prospective areas which were subsequently sold. This system worked reasonably well, particularly in Alberta, where some substantial discoveries of crude oil were made during the existence of the exploration agreement. Unfortunately, in the vast number of cases a discovery was not made, and the chequerboard pattern for leases led to fragmentation of rights throughout the area formerly comprising the permit. This effect tended to discourage other explorers from entering the area, and this system of selecting leases also prompted the need for the drilling reservation, which generally covered that 50 per cent of the land returned to the Crown.

Seemingly, in those parts of the province having relatively easy terrestrial access and where the potential hydrocarbon-bearing formations were not too deep, this system of chequerboard leasing may be satisfactory. However, in areas where access is a major cost and where deep expensive drilling is required, it tended to discourage anyone other than the holder of the 50 per cent leases from entering the area to explore. Certainly in offshore areas where seabed drilling cost is many times greater than land drilling, this system of lease selection is most unsatisfactory.

Leases granted on crown land in British Columbia in 1974 had a primary term of ten years, renewable for further ten-year periods if particular circumstances existed which generally related to production.

The acquisition of leases in British Columbia was handled in 1974 in two separate ways. First was the way previously mentioned, which was the result of having a permit and then earning the right to acquire leases of 50 per cent of the land. The second was through the competitive cash bidding at one of the quarterly sales held by the British Columbia Crown.

There was no obligation at that time on the part of the lessee to drill a well during the initial term of the lease. A 1974 amendment to the Petroleum and Natural Gas Act, however, permitted the minister to forward a notice to drill when he considered that development of the lease was not active enough. This amendment seemed to suggest that before the minister would consider sending a notice, there would, in fact, have been a discovery made on the lease. Thus, in sending the notice, the minister was merely requiring development drilling. It seemed most unlikely that the wording of this section would be interpreted such that the minister could require exploratory drilling on existing leases.

There was an apparent trend to shorten the term of leases, with Saskatchewan granting only five-year primary terms. However, in British Columbia the term of 10 years was not excessive when one considers the short period of four to five months when work can be done in the area of hydrocarbon potential. This is quite different from most of the areas in Alberta and Saskatchewan, where drilling and other geophysical operations can be conducted all the year round.

The government of British Columbia appeared to feel that a large number of leases were held by companies not particularly active in exploring in the province and that something should be done to discourage these holdings, so that lands could be offered to others who were prepared to explore. These feelings may have promoted some of the 1974 changes in the Petroleum and Natural Gas Act, whereby the rental on leases was increased from one dollar an acre to two dollars an acre. The effect of this change seemed likely to be the return to the Crown of some acreage which, under the previous rental of one dollar an acre per year, would have been retained by the lessee. However, one must assume that for the most part relinquished leases would be the least attractive, and leases with potential, even though the lessee was not prepared to conduct any immediate exploration programme, would be retained, and the two dollars an acre would be paid. The greatest effect that this rental of two dollars an acre per year seemed likely to have was to deter other explorers from entering the province. It is one thing to create a system designed to discourage excess holdings by companies not prepared to explore immediately, and quite another thing to expect these same rules to promote the entry of new companies and increase exploration activities.

The royalties on crude oil in British Columbia were essentially the highest in Canada in 1974. The rate was 40 per cent for a well producing 1,000 barrels a month and further escalated to about 58 per cent when the monthly production reached 10,000 barrels. It would seem that this royalty level, which directly affected the producer's cash flow, could only act as a deterrent to increasing the level of exploration in British Columbia much beyond its historical pattern of about fifty to sixty exploration wells a year. Other changes that may be made in regulations designed to encourage greater exploration activity will never perform to their full potential until

the producer's return is comparable with what he can obtain in other jurisdictions.

GOVERNMENT OPTIONS

We can now examine the options available to the British Columbia
government in 1974. First, the government could maintain the existing
regulations without change; or they could have stricter enforcement of the
rules, further reducing cash flows to the producer, making it more difficult
to operate; and thirdly, they could make changes designed to encourage
greater exploration activity in the province. In each of these options it is
assumed that some accommodation could be made between the provincial
and federal authorities with regard to the revenue-sharing aspect. If it could
not, then it would seem almost certain that exploration activity would
decline, not only in British Columbia but also in the rest of Canada, and
that even significant changes in the regulations would fail to act as an
effective inducement to continued exploration.

In the first instance, if the government were to resist changes to its rules
and to continue the existing level of royalty, it could expect to receive about
the same amount of exploratory drilling as there has been in the past. This is
evident from the sale in August 1974, where some 7.2 million dollars was
paid for the right to acquire oil and gas permits, drilling reservations, and
leases. This was certainly an indication that the oil industry was not
prepared to write British Columbia off because of its high royalties and
rather stringent regulations. There are several reasons for this. Companies
have enjoyed a general increase in their cash flow, and in 1974 enough cash
flow was available to pursue exploration in a province where the rates of
royalty would reduce their return on invested capital as compared to other
jurisdictions—for example, Alberta or the United States. The intensive
competitiveness in the oil and gas industry distinguishes it from all other
industries. Companies that have invested a great deal of time and money
developing the geological potential of a particular area and have found that
it fits their exploration parameters are prepared to offer a good deal of
bonus money when this land becomes available through a competitive sale.
Also, because a number of years usually elapses between acquiring an
exploration permit and the development of and production from any
discoveries, the economic climate may change such that what at one time is
not attractive may become economic at the time of production.

If the government of British Columbia wished to discourage continued
exploration by the existing oil and gas explorers and to discourage entry by
newcomers, they could most effectively do this by again raising the rates of
royalty and making the tenure of agreements shorter than they were. This

type of action would cause the companies exploring in the province and those contemplating doing so to restrain their activities; exploration would then stagnate.

Presumably, such action by the British Columbia government would not occur without consideration of these consequences and the recognition that a different vehicle should enter the void left by the existing explorers. This could take the form of a public company, such as the British Columbia Petroleum Corporation; alternatively, the government might feel that, given time, the major oil and gas producers, the fully integrated companies, may finish their exploration for high-reserve reservoirs in other parts of Canada and be prepared to return to British Columbia to look for the remaining reserves which they may need to supply their refineries.

To give any support to this latter proposition, one would have to presuppose that the exploration presently being carried on by the Majors in Canada in the Northwest Territories, in the Arctic Islands, and in the offshore areas of eastern Canada will be unsuccessful and that they will have to lower their sights, accept a lesser prospect, and return. This is very difficult to support, considering the successes already achieved in the Northwest Territories, particularly in the Delta, and in the Arctic Islands. Also, there is a vast geological potential remaining as assessed against the very few wells that have been drilled.

The supposition that a provincial public corporation could enter the exploration field and its endeavours be more beneficial to the province than the present system raises many questions about the corporation's practical efficiency and about the possible political repercussions. This article's purpose is not to examine the political repercussions that might occur when dry holes are drilled with public funds; however, we should examine some of the practical considerations involved in the operation of a public company which has an almost exclusive area within which to explore as a result of discouraging the private sector. For the government to discourage both the existing explorer and the entry of any new ones and to expect the public corporation to be able to fill this void suggests that the government is saying, "If the private companies do it, so can we." Now remember that the private companies comprise all the oil and gas explorers working in the province and those contemplating doing so, given the right opportunities. All these companies have geological staffs, many of whom will be geologists who devote most of their time over a considerable number of years entirely to the study of British Columbia's geology. The public corporation could not expect to have such an extensive source of expertise as that available in the free enterprise system. And so the public corporation would suffer from a reduction in the number of ideas generated.

It is not unusual in oil and gas exploration, with its inherent problem of scientific interpretation and evaluation of geological prospects, to find that one company will acquire a block of land, will explore it, and perhaps even drill on it before deciding the search is unsuccessful. The company will then return the land to its owner, the Crown. This does not mean that there are not any commercial hydrocarbons underlying this land; but rather that that particular company was unable to find them. To find these hydrocarbon deposits, a second, a third, or a fourth company should acquire this land, and, if this is done often enough, the hydrocarbon will be encountered, and production will follow. The problem with the public company being the only explorer in the province is that, unless it is fortunate enough to make the initial discovery, it is very doubtful whether there would be enough enthusiasm to have a second, third, or perhaps even a fourth try at that particular prospect, with the result that the discovery would not be made. This is surely the worst thing that could happen and is probably the most damning argument against a public corporation moving into an area with an almost exclusive right to explore.

If the British Columbia government wishes to increase the level of exploration in the province and so lead, hopefully, to a greater number of discoveries and a better position of self-sufficiency in their own requirements, there are several measures which could be taken to promote such a situation.

The first step that can and must be taken is to increase the cash flow to the producer and to assure the newcomer that if he makes a commercial discovery, he will receive a sufficient return to expand his exploration endeavours in the province. This action should take the form of restructuring the royalty on oil and probably renegotiating the contracts on gas existing between the producer and British Columbia Petroleum Corporation. The 1974 royalty rate on oil was determined by production at the wellhead, with the price not being a factor at all. This means that if the price paid to the producer for a barrel of crude oil should decrease from the present level, then the producer will suffer a decrease in his cash flow, thus his future available financing to continue exploration endeavours in British Columbia will be diminished. A preferred structure on royalty would take into account the possibility of a rising and falling price for the product so that the producer, out of whose cash flows exploratory drilling is carried out, would be the last one to suffer in the case of a decrease in price for both oil or gas. Probably this might best be accomplished by the producer selling his crude oil to the British Columbia Petroleum Corporation in the way gas is sold, but with a better pricing adjustment mechanism than that which exists with the gas contracts.

Another measure that should be taken would be a redesigning of the exploration agreement. This could be done to ensure that the holder could only earn leases after he had conducted actual work on the permit or in the area within reasonable proximity, and that his earnings would be restricted to a consolidated block rather than come from a sprinkling of leases throughout the permit area. This feature alone would help to ensure exploration, because the permit holder must be satisfied that he is getting, at least in his mind, the right half of the permit under lease.

The manner of acquiring oil and gas rights should be reviewed. As mentioned previously, the only method of acquiring an exploration permit under the present rule is through a competitive cash bidding system usually held on a quarterly basis. There are probably several instances where an oil and gas explorer would have been quite prepared to drill wells in British Columbia if he could have acquired the land for a minimal amount rather than having to use money he would put into exploration to purchase land through the cash bidding system. The rentals and fees charged should be the same as in Alberta, if only to appear competitive in this particular area.

CONCLUSION

The exploration for oil and gas in British Columbia cannot be considered in isolation. The proportion of the reserves of oil and gas remaining to be discovered will depend upon the number of exploration dollars allocated to the task. British Columbia's competitive position for these exploration dollars will depend in large measure on its royalty and land tenure policies.

PART FOUR

Proceedings of the Final Seminar

Edited Transcript of the Final Discussion at the British Columbia Policy Conference

CO-CHAIRMEN: ANDREW R. THOMPSON
MICHAEL CROMMELIN

Mr. Thompson: I thought it appropriate to invite Michael Crommelin to size up what he thinks are the most pressing issues that have emerged from our discussion, so that we will have a framework for the discussion that follows.

Mr. Crommelin: I welcome the opportunity to try and define what it is that we have been talking about for the best part of three days, to indicate the issues as I see them that have arisen, to see where there might be glimmerings of agreement, and more importantly to see where there are still substantial differences of opinion.

I think first of all on that list there would have to be this question of rent, economic rent. What is it? Is it a figment of economists' imaginations? Whatever it is, is there any of it in relation to the mining industry and in relation to the petroleum industry?

My understanding is that rent is a surplus. Now, what is the surplus over and above? It is a surplus over and above cost, and when I say cost, I include the normal items that are included in an everyday definition of cost; I include a return on capital, and in that return on capital I mean the return on capital that is appropriate having regard to the particular situation in the industry regarding risk—that is, the return that is necessary to induce the desired amount of capital into that industry. Rent is not a static concept; it is not an amount that is calculated at any point of time without giving thought to a future time. It is the surplus that exists over and above what you have to pay to continue drawing into that economic activity the capital that is required to produce at the rate which society has chosen.

The next very important matter that has come up—it came up very quickly, and it has bothered us since—is the question of risk and/or uncertainty.

A number of things arise here: How much risk is there? What are the opportunities for spreading it? Are the members of the industry risk averters or risk takers? This seems to me to be a critical question not offering a very simple solution. It is a very important question because we have had both sides given to us at the conference, and the implications of the different views are very significant. Does industry dislike risk, or does industry actively like taking risks? Are people willing gamblers or are people prudently conservative in business policy? Now, if industry is a risk taker, that is, it willingly accepts risks, then there is little point in leasing policies that will shift risks over to the government. If industry is averse to risk then consider the further question: Who should bear the risk? Should we try to devise a leasing policy that shifts risks over to the government on the assumption—and I underline *assumption*—that the government is a better entity to bear this risk than private industry?

The next major issue I put before you for discussion is another one that has frequently been referred to, but to my mind not resolved. What are the differences, if any, between what I would call the hardrock mining industry on the one hand and the petroleum industry on the other, that have a bearing on leasing policies? Now, of course, there are all sorts of factual differences between them and I don't want to make little of those, but I do want to ask this group to concentrate on any differences that may have importance to leasing policy. For example, is it a fact that in British Columbia the degree of uncertainty is very different in the hardrock mining industry than it is in the petroleum industry? There have been tentative suggestions that geological uncertainty is very much greater for hardrock mining than it is for petroleum. The second matter of an empirical nature that has been suggested is that there is a difference between the hardrock mining industry and the petroleum industry regarding the size of the economic rent. The suggestion has been made that there is a surplus in the petroleum industry on some pools resulting, I presume, very largely from the diversity in quality among pools, whereas there isn't this great diversity in mineral deposits. In other words, mineral deposits are simply brought into production in the order at which they are required according to movements in prices and the relevant cost functions applicable to the different deposits, and each mineral deposit is marginal when it is brought on stream and remains marginal thereafter.

I have difficulty accepting this position. My inclination is that there must be diversity in quality—there is obviously diversity in location—and that these factors would tend to give you a situation of some mines being intramarginal while other are marginal. The uncertainty associated with the exploration process would tend, I think, to support me in that view, but I invite a good deal of comment on it.

Another matter is the very vexed question of timing. Do different tenure arrangements encourage premature exploration? If so, I think, unless exploration is good in itself, we must assume that it involves a social cost. If there is a necessary link between exploration and development, this cost will be increased. Even if the acquisition of information is good in itself, perhaps there should be a clear break in the tenure arrangements between exploration and production: that is, exploration rights should not necessarily lead to production rights. Now, of course, this leads to all sorts of questions about how you should set up your tenure arrangements. I am not suggesting that mining companies would be very willing to go out and explore for the benefit of adding to mankind's knowledge, but that does not mean that you cannot have exploration carried out by somebody adequately remunerated without tying exploration to development rights.

Finally, there is the question of public enterprise. In the discussions that came up at the conference, it seemed to me that the only course for public enterprise was considered to be one of public enterprise monopoly. We didn't consider the question as one of degree. Anthony Scott at one point raised the question, directed, I think, to Frederick Peterson, "Are you in favour of the geological survey in the United States? If so, are you in favour of the present size of the geological survey budget? If not, is that budget too small or too large?"

Now, it seems to me that this is a matter of degree. You may be prepared to suggest that there is a role for public enterprise without going to any of the lengths criticized by Arlon Tussing. In some particular circumstances, such as the very early stages of exploration, there may even be an argument for public enterprise monopoly; I do not want to rule that out. But I am saying that public enterprise can and should be discussed in a broader context. It is a question of degree, and it is a question of the cost of alternatives. May some of the faults inherent—and we have noted them—in private leasing arrangements be solved by the introduction of a public corporation at some stage? If so, what sort of activities lend themselves to this? Arlon Tussing told us what the public enterprise should not do. But he didn't tell us what it may be able to do and what, perhaps it may be able to do more efficiently than private enterprise.

Mr. Mead: The cochairman identified the first major thing that we are talking about here to be economic rent. There was a lot of pretty strong commitment among the economists here that the role of government or the objective of government ought to be to collect the economic rent. I would suggest one modification of it (or "emphasis" I think is the more correct word), that is, there ought to be economic rent in the social sense, not necessarily the gains accruing to private firms. But once you have said that government is entitled to economic rent, then that leaves some

major chores ahead. How are you going to collect it? I think the problems
of oil and gas on the one hand and hard minerals or, say, timber, on
the other are really quite different. John Helliwell's excellent article
identifies a great many alternatives that we all ought to be aware of because it is not enough to say "collect economic rent." John Helliwell
reminds us that there are a lot of ways to do it. And whatever way you
choose, the consequences are different. They are different because they
may affect the size of the pie. If the method of collection reduces the size
of the pie then everybody loses. I think it ought to be a second objective
of government to not reduce the size of the pie. Its method of collection,
in other words, must be efficient. And, for example, there is this matter
of using actual costs. As soon as you do that you are immediately
confronted with the public utility problem. You are in effect saying to
every company, "You may have a fair rate of return on your money, and
that is all," which completely destroys any incentive for that firm to be
efficient, and the pie goes down. It is a method that you just can't use.

I believe that Anthony Clunies Ross in his proposal for the resource
rent tax has this same problem. As soon as you say to a company, "You
are limited to an 'x' rate of return," whatever that is, you are guar-
anteeing that it will not be efficient.

One of the very attractive features of competitive bidding that intrigues
me is that it avoids these problems. If you have reliable competition—
if needs to be underlined—then, in effect, you invite people to come in
and bid. If competition is effective, then they will bid away the
economic rent—not in the particular sense but in the aggregate. I think
that is what this study that I did shows. It shows that in the aggregate the
government collects at least all of the economic rent and maybe a little
more, which left the possibility for firm A, which is very efficient, to get
rich and firm X to lose its shirt if it was not efficient. But in the aggregate,
if competition is effective, bidding will collect economic rents: it avoids
the problem of using actual costs; it avoids this public utility problem;
and it avoids a lot of the problems of assessment and appraisal. However,
if competition is not effective, then one needs to ask what you can do to
protect the public interest. In timber they have used appraisals frequently
to protect the public interest. But another thing you can do is to use
sealed bidding, so even though you do not have a large number of
competitiors, you may have the effect of competition.

It seems to me that also there is another method that Canada ought to
consider and that is the profit share bid; not fixing the profit share—
which I think is what Clunies Ross has in mind—but profit share bidding.
Rules have to be set down, specific rules for how profits are to be calculated. Everybody has to know in advance what those rules are. It may

not make an awful lot of difference what the rules are. For example, you may say company overhead is not going to be permitted as an expense. That is no problem, as long as the rule is known before the bidding takes place. If it is clear that company overhead is not going to be included as an expense, then the calculated profit is going to be shown as a little higher, which will mean that the company in making profit share bids will bid a little lower.

And then one other point, it is not fair to change the rule in mid-stream. You have got to stick with it. If you change it in midstream then you are doing something that would have led that company to bid less or more. This rule is violated all the time. For example, on the big oil shale lease sales that we have just had in the United States, I recall that in the first one, a company bid over $200 million. There was an interesting little clause in that thing that said we are going to divide the 200 and some million into five equal payments, then it says we may never collect the last two fifths of it. Nobody knows whether they are going to collect or not. So, what does the company do that comes in and wants to bid knowing that the rules are uncertain? The government is probably giving away something without telling it. Politically, I suspect that the government cannot at a later date decide to collect that fourth and fifth share of the bid. If they cannot do it and do not do it, then they have probably given up something, because companies probably bid a little less thinking that they might collect it later on.

Mr. Ross: I felt that Walter Mead was misrepresenting what I said in that our proposal quite explicitly recognized that if you attempted this rate of return threshold you could not afford to collect all those returns above that level. I thought he was rather representing us as saying that one allows so much rate of return and no more. I think we did quite explicitly recognize that this was not a possible position to take.

Mr. Mead: What was the percentage of the profit that you would take beyond the threshold point?

Mr. Ross: Well, in the examples we used a first threshold beyond which we take 50 per cent and a second threshold beyond which we take 75 per cent, and I think in the course of the article 80 per cent may have been mentioned at one point, but the examples are framed in those terms.

Now, I thought you had raised earlier another objection to this, that even if the rate were only 80 per cent or 75 per cent of the maximum, this was so close to 100 per cent that there would be certain wasteful expenditures.

Mr. Mead: Yes, that is the way I would make my objection.

Mr. Ross: I think it has been mentioned that, in principle, if one could really try and be truthful with the expenditures and the receipts, there need be no incentive for wasteful expenditures in that situation. I just wondered whether your assertion about this was an empirically based one, and if so, at what marginal rate these wasteful expenditures begin to operate.

Mr. Mead: Well, it is based pretty largely on the experience that we had in the United States during World War II, where we had an excess profit tax rate of 85 per cent. What we learned was that, since you kept only 15 per cent of whatever you saved, it wasn't much of an incentive to operate efficiently.

I can't tell you what the cut-off point is. It is somewhere in that area. Somewhere between, I suppose, 50 and 85 per cent.

Mr. Ross: I suppose really the point that I want to make is that there is no *a priori* reason for this; it is only because of imperfections in the accounting system.

Mr. Mead: No, not at all. It refers to the decision making structure and how businessmen make decisions. If you tell them that we are going to tax away 85 per cent of their earnings on an incremental basis and you get into that 85 per cent bracket, it leads them to be very careless about economizing.

Mr. Ross: I think with the profit sharing approach, inevitably the government can collect only part of the rent. With the cash bidding approach, in principle, it can collect the whole of the rent with sufficient competition.

But I thought the point was worth making that in the case of risk aversion on the part of the private companies, the total amount of the rent that you may be able to collect with cash bidding may be less than the fraction of the rent that you can collect with the profit sharing. This depends upon the assumption of risk aversion.

Mr. Mead: It might very well be. And in that sense it is an apples and oranges comparison. Because one, in my view, should not use a profit share bid where things are quite certain. The profit share bid has merit in a situation where there is great uncertainty, where you don't know what the costs and the benefits are going to be. And it has merit in the mining situation where uncertainty is uncommonly high.

Mr. Gaffney: Mr. Mead, would you consider splitting it up vertically? How would you react to the proposal made earlier that exploration up to a certain stage be contracted out by the Sovereign and then, that having been done, the resource could be put up for bid.

Mr. Mead: I think that would be a very reasonable thing to do. The Sovereign bears the cost of exploration. And then, at the point at which the Sovereign has identified minerals, the Sovereign might very well ask for bids. Uncertainty has been reduced, you know more about what you have got, and I should think that mining companies at that point could come in and make a bid for what exists; and because uncertainty has been reduced, one might at that point go the bonus round, or you might go the profit sharing round, either one.

Mr. Gaffney: Would you also consider it desirable at that stage to let the bonus be paid in annual installments over five or ten years?

Mr. Mead: Yes, I think that is worth thinking about. I see only one objection to it, and that was brought out in the discussion; if you delay a bonus, that lets a firm find out whether or not they have been taken in the process. If they have, some firms will elect to go the bankruptcy route and not pay a bonus. Some firms cannot afford to do that because they are in business forever. If that is the case, it puts the more responsible bidding firm at a disadvantage with respect to the less responsible.

Mr. Ross: If you went for profit sharing because of a lack of competition, what sort of profit sharing arrangements would you go for?

Mr. Mead: Well, in a hardrock mining situation, I would be inclined to permit all costs to be deducted that are clearly attributable to the mining venture but not company overhead—that is a minor item anyhow. And I say that because I learn from mining people that there are great differences between mines owing to the characteristics of mines in terms of how much capital you are going to be investing. So you might deduct all capital charges in computing profit. The rules for deduction, of course, must be clearly specified.

Mr. Thompson: Are there not differences in the technologies which different mining companies might employ on the same orebody or different kinds of capital structures and, therefore, efficiencies and inefficiencies in the decisions that they may make; if so, the arguments used a moment ago against using actual costs would seem to apply there. How do you reward the man that makes the optimal decision about the way, the amount of capital, and the technology to develop the mine.

Mr. Owens: I don't think this is a problem, because once you get to the point of having a mine, almost every company would go about its development in the same way. One might make it a little bigger than another, but, by and large, economics will determine the size. Preliminary exploration is a different situation, and so might be intermediate stages of exploration. One company might spend quite a bit; another company might spend very much less.

Mr. Mackenzie: If we are to consider the profit share bidding and look at it in relation to hardrock mining, one of the essential things is at what stage it would be practical to apply it. Say we agreed that it might not be practical at the early exploration stage, but it would be a more practical proposition to consider at the development stage. Then the question I am asking is "what would you do about rewarding exploration, if you wanted to leave exploration in the private sector?" Perhaps you would have to reserve a share of the profit for the explorer and then bid on dividing the rest of it up.

Say—it was Mason Gaffney's idea—that you reserve exploration for the Sovereign, and that it could be done on a contract system. I think the particular problem there would be how to motivate the Sovereign effectively as far as exploration is concerned, because exploration isn't a matter of routine, it is a matter of applying some unusual concepts and skills. You have to have some sort of effective reward system to be able to motivate that. It is not just the Sovereign falling into money as a matter of routine and discoveries turning up that could be put out for profit share bid to be picked up by mining companies that would develop and operate them. Those are some of the problems I see.

Mr. Mead: It may be true that the Sovereign can not efficiently simply contract the job out. If that is the case, then you might want to ask for the profit share bid at one prior step, namely, before the area has been searched and explored.

Mr. Mackenzie: If it is done that way, say at the early exploration stage, isn't it a possibility that the potential value of an area would be so uncertain at that stage that it would be hard to get meaningful bids?

Mr. Mead: No, no, no. That is precisely the purpose of the profit share. And are there firms in existence that other private firms can hire to explore for minerals?

Mr. Mackenzie: Yes.

Mr. Mead: Why cannot those firms be hired by the Sovereign to explore for minerals?

Mr. Mackenzie: When an exploration organization contracts out to somebody else to do the exploration, it wouldn't normally be in terms of carrying out the whole exploration programme, but only for doing a particular survey or carrying out a particular drilling programme, or something like that.

Mr. Mead: So they can be hired to do a particular part of it; other firms can be hired to do another part. Is that true?

Mr. Mackenzie: Yes. The routine aspect can be contracted out, but I don't think the decision can be. I think that is important because that is where the concept and skills and the quality come into play.

Mr. Mead: So the government would have to perform that decision making function itself, the final decision making job.

Mr. Mackenzie: If it didn't, it could only effectively contract out on some sort of a profit sharing basis to the exploration group to motivate that aspect.

Mr. Wedge: One of the things that we haven't really examined is the problem of timing. Whether you lease at the beginning before exploration is done or whether somebody in government decides on exploration and then leases when you have got some sort of information, you have got a bureaucratic decision to make about the timing of when to do exploration or when to lease a specific block of land.

We have talked a little bit about the diversity of firms and geological ideas in the industry and about how that applies to the impossibility of one monopoly doing all the exploration. There is general agreement among the people who have talked about it that there is a great value in diversity of opinion and the diversity of people trying different approaches that you don't get in a monopoly. That applies just as much to the monopolist who decides when to release some blocks of land for exploration as to the monopolist who does the initial exploration and lets it out to private bids at a specified point.

Mr. Thompson: What is an alternative?

Mr. Wedge: The alternative is the plain system where a firm goes and takes a block of land where it is worth the cost of filing the claim.

Mr. Thompson: Will you approach it in this way in answering, and it might help me to figure it out: Suppose that the Hudson's Bay Company had all that land—not the Crown but the Hudson's Bay Company or the Canadian Pacific Railway. How would they approach it? Would they simply put their stock on the shelf any time anybody wanted to buy, or would they figure that there were time preferences in terms of their introducing the stock to the market?

Mr. Wedge: If I were selling apples on the street corner that is the way I would do it.

Mr. Owens: They have put some of their land out on a profit sharing basis.

Mr. Thompson: Oh, sure, I agree, they have put them out. But they don't simply put them out because somebody says put them all out so that we can bid.

Mr. Gaffney: Well, just on the apples-on-street-corner thesis, it seems to me that people who produce apples put them into cold storage and spread them out over the year, because the price goes up as they become increasingly scarce. Speculation is a useful function, and because it is useful it has an economic reward. Andrew Thompson's point is that it is desirable for the Hudson's Bay Company to hang on to this land until its price has gone up.

Mr. Wedge: Well, if they don't know a lot about the industry, that is the approach that they may take. They may have an unreasonable idea of the value of their property. That is a common problem with people who have mining rights.

Mr. Gaffney: Well, if they do know a lot about the mineral industry, what would be the next step? That is what we are here to find out.

Mr. Wedge: They do their own exploration.

Mr. Gaffney: Well, is that what you want the Crown to do?

Mr. Thompson: . . . and then market it on the basis of what they turn up?

Mr. McPherson: Not market it. The Canadian Pacific Railway is not marketing it. They would do their own exploration and go into production.

Mr. Crommelin: All right, take the example of the Crown having all the mining rights, with just a few exceptions, in British Columbia. What does the Crown do? Does the Crown institute or at least continue the system of claimstaking that has been strongly defended recently in British Columbia by at least some sections of the mining industry as the best method of allocating mineral rights to the industry; or does it not only try and influence the rate at which the mineral rights are allocated, and does it sell the mineral rights?

Mr. Owens: Well, that is a pretty big question.

Mr. Crommelin: What I want to know is what is the case, what is the argument in favour of the free entry system.

Mr. Owens: Isn't it really dependent on what the public interest is at that particular time?

Mr. Crommelin: Yes.

Mr. Owens: And the Crown has traditionally been in the position of acquiring vast lands—initially it didn't even know how vast because it was in London, and later it got a little bit closer to the scene and it still wasn't sure, and it certainly wasn't sure what was on the land. As its public became interested in it, the Crown devised a system of parcelling it out. Initially, whether it was land or minerals, you put a post on it and you were granted the rights, they were yours.

Now, as more is known about the land and more is known about the public need, the Crown has modified the system and it has brought in other forms of mineral tenure. Presumably that responds to the need of the moment.

If you want to stimulate activity, surely you bring incentive schemes in, or you do preliminary work to show that the ground has attraction. If you are not concerned with that, why not leave it for the individual to acquire the land under whatever terms—and they can be as favourable as possible in that the total government takes will always be 50 per cent at least, because that is the gross taxation rate.

Mr. Mead: Excuse me. Fifty per cent of what?

Mr. Owens: Fifty per cent of the profits, of earned income.

Mr. Wedge: That is an advantage that the Hudson's Bay Company doesn't have. The Hudson's Bay Company doesn't get that tax paid automatically.

Mr. Crommelin: Oh, yes, I recognize that. But what is the argument that says if you open it up and allow free entry, this serves the public need? It is not obvious to me that the mining industry's rate at which it stakes claims is the same rate that suits British Columbia's society best. That doesn't follow.

Mr. McPherson: And the mining industry takes the reverse side; you haven't proved that the development of the mining industry to date has been harmful to the people.

Mr. Crommelin: Well, I suggest that there is a reason why the rate at which the mining industry goes in is faster than the rate that is appropriate for society. That one reason is that all members of the mining industry know that as soon as there are any prospects of any particular area being worth something in mineral terms, if they don't get a piece of the action then someone else will. In other words, you are placed in a competitive situation in the acquisition of land rights, and the competition focuses on the point of time at which you enter, stake, and register. So the competitive situation under which you are placed forces you to go in faster than perhaps you would if you knew that you could wait a little longer and the stuff might still be available to you.

Mr. Wedge: The cost of staking claims is so small in relation to the total cost of exploration that it is very difficult to argue that you are dissipating any significant amount of money by allowing claimstaking rushes, or whatever.

Mr. Crommelin: Let's assume that the cost of staking a claim is absolutely free. My argument is that the mining industry would stake faster than the socially desired rate.

Mr. Wedge: Mr. Crommelin, if the cost of staking a claim is zero, and it is all staked today and nobody does any exploration work, how does the fact that the claims have been staked faster harm the people of British Columbia?

Mr. Crommelin: I am prepared to concede that if it is absolutely costless to stake claims and maintain them, then there will be no inefficiency in a free entry system. Doesn't it follow that if it costs you as much as one cent to do so, that the one cent is inefficiently spent?

Mr. Wedge: Not necessarily, if the cost of staking a claim is minimal. What you are really concerned about is whether the exploration is worthwhile. You stake a claim when you think it is worthwhile to do exploration. You don't go out and spend the cost of staking a claim unless you have some particular reason for going out and looking at that ground.

Mr. Peterson: I am going to give two other reasons why I think claim-staking is bad. I think one reason is that if that land is worth anything, then some rent may have gone by the boards. If somebody doesn't have any right to it, then maybe it ought to go to the state.

The second reason is that once this claimstaking has taken place, presumably the ownership is so fragmented that if someone later wanted to go in and really pick up a big tract, they would have an awfully hard time negotiating with all these people and putting the block together.

Mr. Owens: I can accept Michael Crommelin's point. I think the only thing is that the relative amount of money in the metallic industry is very insignificant. In oil and gas it might be higher.

Mr. Bradley: I may be taking an oversimplified point of view, but I am suggesting that if you design a policy, the net effect of which is a level of taxation from the industry corresponding to the rate of growth that you would like to see in that industry, that is the one control that you have to be worried about. You do not have to have somebody in the provincial government trying to outguess the mining company management as to when is the best time and what is the best scale to go forward on. That would be a kind of a fine control that doesn't appeal a great deal to me personally.

Mr. Gaffney: It seems to me that Michael Crommelin is concerned about the economic waste, claimstaking and the economic waste. Well, let me say first of all that I do not understand this claimstaking process nearly as well as I would like to; that is one of the things that I am hoping to learn from listening to you people. But, to the extent that I do understand it, the waste would come in where the Hudson's Bay Company or the Province of British Columbia is giving away this resource at a time when it could start getting money for it, when, if it held on to it for a while longer, it could sell it for a good price.

Secondly, if you have a work requirement, this becomes another kind of loss—a phoney work requirement—and, as you know, an awful lot of this sort of thing goes on in various countries of the world.

It has been stated several times that the amount of money that goes into exploration is extremely small. If that is true, then we are talking about peanuts, and it doesn't really matter anyway. However, I think that several billion dollars a year go into the search for oil and gas. I don't know what the effects are in hardrock.

Mr. Thompson: There is a related subject here that we might explore. There has been a lot of criticism as to the discretion that is exercised at the time that it is decided whether or not a mine can go ahead. Now, I want to leave aside for the moment the issue of whether the decision is too arbitrary and not fixed, with adequate safeguards and standards and what not. What I am really asking here is, "What is the position of the policy maker, if you assume that there are severe social costs—they may be offset by benefits—involved in the opening of any mine. I am referring to the fact that there are going to be demands for improved roads, townsites, schooling, and hospitalization, and other factors which impose quite severe costs on society. Should there be a component of public policy dealing with whether or not that mine should go ahead at that point, bearing in mind these costs? Secondly, if there should be such a policy how should the decision be made?

A mine may be a profitable venture to the miner, and yet it may impose a tremendous net deficit on society. This is a possibility that we have to face. Nevertheless, it is suggested that the decision should be exclusively the miner's decision and not government's.

Mr. Wedge: You forgot to take Paul Bradley's approach in deciding whether the net cost of providing the infrastructure is greater or less than the net increase in the tax revenue which is likely to result if you survive.

Mr. Thompson: So your answer to me is, to have this kind of analysis, and, on the basis of it, you decide yes, the mine can go ahead, or no, it can't.

Mr. Wedge: I would say you can decide yes, you will provide the schools or no, you will not provide the schools, and, if the mine wants to go ahead, then it will have to provide them.

Mr. Thompson: Well, then you would have to lay down conditions for a security bond, as they lay down conditions for coal mining to rehabilitate the land. There is no way that society will permit any government to pass off the responsibility to provide education to a private company; that is just impossible.

Mr. Wedge: There are mining towns where the mines have built roads, hospitals, schools, and most of the social overhead capital, and it is a matter of negotiation with the government whether or not that capital is provided.

Mr. Owens: I think I would approach this in a little different way. I would say that the Crown and the miner are in this thing together, basically. They are going to share the profits in some way approaching equality. The person who is going to work in that mine is going to live somewhere, and he's going to send his children to school somewhere. There is shortage of housing in Vancouver, and it really doesn't matter a great deal whether you build the housing in Williams Lake or in Vancouver. In fact, it may be cheaper to build it in Williams Lake because land is cheaper. So that really I am not sure that these considerations that have been raised are all that important to this decision. It seems to me that the taxing structure has already been designed to take care of that.

Mr. Crommelin: Can I ask you a question on that. If the government and the mining company are, as you say, in it together, which, under the circumstances, I certainly agree they are, doesn't it follow that at the very least the decision should be taken between the two of them? It doesn't seem to follow that, because the government and the mining company are in it together, the sole rights for deciding whether production will take place should lie with the mining company.

Mr. Owens: Consider the practical situation. Suppose that a mine is found. Now, there is the problem of the production plant and the income structure. The company is going to look at the income structure question and is going to talk to the government and say, look, you have built houses in Ottawa, how about some Central Housing and Mortgage money for houses in this town?

So, I think, it naturally works out as a joint decision anyway. I think the discretionary opportunity exists in the natural situation without it being applied specifically and only to the mineral laws.

Mr. Thompson: The question that I want to ask relates to a matter that we discussed earlier. The profit share bidding that you were suggesting, Mr. Mead, has been used in the oil situation in Saskatchewan and has also been attended by very great difficulties, and I think it has been used in the United States.

Mr. Mead: Well, profit share bidding has been used for the offshore from Long Beach, and the people who administer it are very happy with it. In my opinion it was a disaster. However, it need not be.

Mr. Thompson: There is such a tendency for people to bid at a profit rate that is unrealistic.

Mr. Mead: But you can arrange for firms not to bid at a high profit rate by a simple device. If you want to keep the profit share down to, say, 25 per cent, there is a very simple way of doing it, and that is, to request a fixed bonus payment which is not deductible from the profits. Then firms will bid an amount which will enable them to recover the profit before the profit sharing, and it will keep the profit share bid down.

Mr. Thompson: But even if you had that requirement, isn't there still a tendency to overbid?

Mr. Mead: Well, if that is the case, that is industry's business.

Mr. Thompson: But isn't the result too frequently that as soon as it appears that it is nonprofitable, that the nature of the discovery is such that there is no way that you can make it go, there is likely to be abandonment, or more than likely what happens—and this is what we have seen—is that they will come back to the government and try to lower the profit share.

Mr. Mead: No, I don't think so. That is one of the nice things about a profit rate. Suppose somebody overbids. Suppose he should have bid 25 per cent and he bid 50 percent. Well, that still leaves 50 percent. You can't go bankrupt. You have still got 50 percent of the profit. There is no way to go bankrupt. So, it is not a problem.

Now, the bid itself is not a problem. If you are so inefficient that there is no profit, then of course it doesn't make any difference what you bid.

Mr. Ross: If it is worked out before you pay off the debt, you could go bankrupt, couldn't you?

Mr. Crommelin: If you borrowed at 15 percent and your 50 percent share of the profits was in fact returning you 12 percent, you would go under, wouldn't you?

Mr. Mead: Yes, I suppose so. That is a possibility.

Mr. Owens: There was just one point that I thought I would make, and I don't expect to convince everybody, but I would still like to be convinced the other way around on this score at some time, some place, not necessarily today. We discussed profit sharing. The assumption has sometimes been made that all the economic rent, however that is defined, should accrue to the landlord. However, I detect through the conversation that really we are talking about only a certain percentage, and somebody put an upper limit at 85 per cent, although I am not sure that many people would support that number.

Mr. Mead: That 85 per cent discussion didn't concern economic rent; it concerned profits. You are mixing the two things.

Mr. Owens: I still believe that nobody has made a case—because nobody has tried to make a case, of course—that really the government should collect the rent at all, and certainly, anyway, not that they should collect it all. It is not obvious that government collecting it really satisfies the social needs better than if the government is somewhat inefficient in its collection. Really, in the long term, what is important is how quickly the rent is going to get into increasing the productive capacities of society. Government is one way, and it really hasn't necessarily been established to all people's satisfaction that it is the most efficient way.

Mr. Crommelin: Could I just briefly answer two points there. One is, I think, you have a problem of efficiency if you are in a country or a province like British Columbia, where a proportion of your investment capital may be foreign capital. I think, as someone pointed out, if the government does not collect the economic rent, then it goes to the foreign entrepreneur. So I think that is an excellent argument for the government collecting the rent.

Mr. Owens: Yes.

Mr. Crommelin: The second thing is that as a member of, say, British Columbia's society, I would be concerned about the distribution question; that is, if this province owns mineral assets worth x dollars, I wouldn't be altogether satisfied with the thought that that x dollars would

be shared out amongst the mining companies active in the province. I
would think that in some way it belongs to the province as a whole, and
the only way I would know it was getting to the province as a whole
would be for it to go to the province's representative, the government.

Mr. Owens: I was only talking about efficiency, and, of course, as long as
even the foreign money is ploughed back in and not exported, there is no
problem there.

Mr. Gaffney: It is a complete red herring to assume that an increase in taxes
on natural resources is matched by an equal increase in government
expenditures. The two things are of separate concern. I will agree that
certain government expenditures are too high and I agree that certain
taxes are too high. It is a question of how the tax burden should be
distributed.

Mr. Thompson: I think we have wound this down. Thank you all, it has
been very interesting. On that note this conference is completed. Thank
you.

Concluding Note: Economic Rent and Government Objectives

MICHAEL CROMMELIN

Most prominent amongst the objectives advocated for a government system of mineral leasing is the collection of an "economic rent" derived from mineral production. This is strongly urged by Mason Gaffney in his opening article and is supported (with greater or lesser emphasis) by Walter Mead, Anthony Clunies Ross, and Brian Mackenzie. On the other hand, Jim McPherson and Owen Owens suggest that any attempt to apply the concept of economic rent to the mineral industry involves a basic misconception. Paul Bradley feels that classical rent theory applies only in the short term, whereas a government devising a mineral leasing system should be more concerned with the long run. He therefore proposes government revenue, expressed in terms of present value, as a more suitable objective. David Quirin and Basil Kalymon also discard economic rent in favour of maximization of social welfare. It is obvious from discusssion on this point that not all parties define the concept of economic rent in the same way. A brief review of historical usage appears warranted.

Adam Smith regarded rent as "the price paid for the use of land." He noted that it varied both with relative fertility and with the location of the land. Moreover, land could produce different amounts of rent according to its use. One possible use was mining. Not all mines would produce rent; whether a mine did so depended partly upon its fertility and partly upon its location, as with land in general.[1]

Malthus emphasized that rent consituted a surplus. He defined it as "that portion of the value of the whole produce which remains to the owner of land, after all the outgoings belonging to its cultivation, of whatever kind, have been paid, including the profits of the capital employed, estimated according to the usual and ordinary rate of the profits of agricultural capital at the time being."[2] He was not concerned whether the concept could be extended beyond agriculture to mining.

Ricardo was more precise. As the passage quoted by Ross Garnaut and Anthony Clunies Ross at the beginning of their article clearly shows, Ricardo recognized that rent was a function of differences in quality. The marginal mine produced no rent, but simply returned the costs of production, including normal return on capital. Mines of superior quality (whether this resulted from the nature, grade, or location of the deposit) returned all these costs plus something more, the additional amount constituting rent.[3] However, while Ricardo expressly applied his view of rent to mining, he also said that "rent is that portion of the produce of the earth which is paid to the landlord for the use of the original and indestructible powers of the soil."[4] This narrow definition seems impossible to reconcile with the general approach based on quality differences.

John Stuart Mill pointed to the fact that "mines of different degrees of richness are in operation, and since the value of the produce must be proportional to the cost of production at the worst mine (fertility and situation taken together), it is more than proportional to that of the best. All mines superior in produce to the worst actually worked, will yield, therefore, a rent equal to the excess."[5]

Alfred Marshall was apparently influenced by Ricardo's narrow definition in suggesting that mining be treated differently from agriculture when defining rent, because mineral deposits are exhaustible. Thus, the excess of mining income over outgoings had to be regarded, in part at least, as the price paid upon the sale of stored-up goods. Rent, it seems, was exclusive of this price.[6]

Joan Robinson adopted a more general approach. She cast off the assumption implicit in the writings of some early economists that land had only one use. Her definition of rent is worthy of quotation:

> The essence of the conception of *rent* is the conception of a surplus earned by a particular factor of production over and above the minimum earnings necessary to induce it to do its work. This conception of rent, both verbally and historically, is closely connected with the conception of "free gifts of nature." The chief of these free gifts of nature (of which the essential character is that they do not owe their origin to human effort) is space, and for this reason they have usually been referred to simply as "land"—land being understood to comprise all the other "free gifts" besides mere space. Consequently the term rent, which in ordinary speech means a payment made for the hire of land, was borrowed by the economists as the title of the sort of surplus earnings which the free gifts of nature receive. The whole of the earnings of *land* in the economist's sense is *rent* in the economist's sense, for it follows from the definition of the free gifts of nature that they are there in any case, and do not require to be paid in order to exist.[7]

It is this fact that rent amounts to a surplus that has proved so enticing to economists. The reason is to be found in the hope that a government may be able to collect this surplus without affecting the level of activity in the industry. Classical economic theory has long suggested that enjoyment of the surplus is a matter of distribution rather than efficiency. In other words, private operators in the mining industry are influenced by marginal returns and costs rather than total returns and costs. Willingness to invest is reflected in the supply price of capital which, by definition, is paid to the operator before rent is derived. There is no doubt that the operator would prefer to receive any rent in addition to all his costs, but this is not a *necessary* condition for obtaining his investment. Furthermore, payment of some or all of the rent to the operator will not induce more investment because this has already been fixed at the most profitable level.

However, this discussion of rent provides merely a beginning, not an end, to a government's search for policy objectives regarding mineral development. A number of further problems complicate this simple picture almost beyond recognition.

A generalized theory of rent may well be concerned with social revenues and social costs, the latter including an element of opportunity cost representing the revenues that could have been produced by use of the land in the next best industry. However, a government involved in the management of publicly owned mineral resources must look not to this figure but rather to the difference between *private* revenues derived from mineral development and *private* costs incurred therein.[8] To do otherwise would be inconsistent with the fact of public ownership of the resources. For example, if a block of publicly owned land has two possible uses, mining and recreation, social rent will be less than private rent. On the assumption that mining is the higher use, social rent will be the difference between social revenues and costs including the social opportunity cost of the land used for recreation; private rent will be the difference between private revenues and costs, which do not include any such opportunity cost. Private rent will therefore be larger. To induce mining on the land, the government is not obliged to compensate the private operator for the use which is thereby precluded because the land is publicly owned. It is the public who must be compensated. Thus, the government must seek to capture the rent calculated in terms of private rather than social revenues and costs.

Social benefits and costs are relevant in another context. They should be considered by a government in deciding whether to allow mining in any particular area. The existence of rent, defined as above in terms of private revenues and costs, does not determine this issue. For mining to be allowed, there should be a net social benefit. The opportunity cost of land is an important element in this calculation.

It is sometimes suggested that mining operators, if they are to continue in business beyond the life of a particular deposit, should be compensated not only for the costs incurred in production but also for the value of the depleting assets. This idea may have its origins in the United States, where a significant proportion of mineral resources is subject to private ownership. However, in a jurisdiction where the government is the owner of minerals in the ground, it is the government and not the private operator who should obtain compensation for the asset. To pay the operator more than his costs of production would amount to a free transfer of the publicly owned asset to the operator, in addition to the return of all his costs. Similarly, the position is sometimes taken that, since it is necessary over time to develop progressively lower grades of mineral deposits (the better grades having been developed first), mining costs must inevitably rise and a private operator must recover *replacement* costs as distinct from *incurred* costs in order to stay in business. Even if rising costs were inevitable (and this ignores the impact of improved mining technology), the suggested result would not follow. Each new mine should be judged on its own merits, namely, whether it is capable of returning the supply price of capital to the operator, and whether it will produce a net social benefit, without regard to previous cost trends. If it fails on these grounds, society would not wish to undertake it; if it succeeds, it should be capable of attracting the necessary labour and capital for development.

The selling price of minerals is most important in determining the size of the surplus that has been defined as rent, since it determines the revenues received by the producer from which costs are deducted. However, it has sometimes been claimed that the opposite is true, namely, that the amount of the rent extracted by a landlord determines the product price. This would be the case only where demand for the product was completely inelastic, that is, where the amount of product purchased was completely independent of its price. This may occur with some products during temporary shortages, but such occasions are likely to be rare and short lived. The normal circumstance facing a government, particularly that of a province, is one of price determined by such external forces as world supply and demand for the mineral. The government must accept this price and its influence upon rent. If it seeks to impose charges upon a mineral producer amounting to more than the rent calculated by reference to this price, production will be curtailed.

It is also important to distinguish this concept of rent from taxation. A government may impose a variety of taxes upon a mineral producer as a means of collecting some or all of the available rent. However, such taxes are different in principle from the general taxes imposed on industry and individuals in order to pay for government services. General income tax is the most significant type of payment in the latter category. This tax is designed to operate as a charge upon all income-producing activities conducted within the jurisdiction, quite apart from any government contribution to that activity. Taxes aimed at rent collection aim to compensate the government for the publicly owned asset which it has contributed to the undertaking. Accordingly, rent collection and general taxation should be viewed separately. This does not imply that the two activities are unrelated—John Helliwell illustrates the contrary in his article in this volume. The relationship is of particular significance in a federation where one level of government is the resource owner and the other has control over general taxation. The level of general taxation, and particularly any provisions applying to special industries, such as mining, affects the size of the rent available for collection by the owner. However, the fact that mineral producers are taxed in different capacities should not give rise to any plea for relief from "double" taxation, for the collection of rent is a separate function from ordinary taxation.

The essential problem associated with rent collection as a policy objective in mineral leasing concerns the identification of rent. If a government waits until a mineral deposit is exhausted before calculating rent so that it is in a position of knowing actual revenues and costs, the collection of rent may be deferred for many years. Even then, the government has the problem of ascertaining the supply price of capital to the project. Moreover, there is little incentive for the private operator to minimize costs if all charges are allowed by way of deduction from revenues; but if all charges are not allowed, there is the administrative nightmare of deciding which costs are deductible and which are not. If a government is not prepared to wait until the conclusion of a mining operation, rent must be calculated having regard to *expected* revenues and costs. These must be estimated before all the characteristics of the mineral deposit are fully known, thus introducing the element of risk. If the quality of a deposit turns out to be lower than expected, or if price falls below what was anticipated, the private producer may suffer a loss. On the other hand, if either quality or price is more favourable than anticipated, the government will not succeed in collecting all of the actual rent. The political repercussions of this alternative may be severe.

Identification of rent is not the only problem. Considerable significance is attached to the method by which a government effects collection, for different methods involve different collection costs. If a government adopts a system of cash bonus bidding, as suggested by Walter Mead in his article in this volume, the cash bonus represents a sunken cost which has no bearing on subsequent development and production decisions based on considerations of marginal revenue and cost. The result is therefore "efficient" in the traditional sense employed by economists. However, under this method, the rent is paid at a very early stage of operations, and this gives rise to the question whether the time preference of a private operator coincides with that of the government. In other words, is the discount rate employed by private operators in reducing future revenues and costs to present values (quite apart from any allowance made for risk bearing) the same rate the government would use? If there are indications that the government's rate would be lower, there would be an advantage for the government in substituting rent payments made throughout the operation for a single sum paid at the outset. Another popular method of rent collection involves a royalty paid as a percentage of production as it takes place. If the royalty is calculated without any deduction made for costs, it is known as a *gross* royalty, whereas, if deductions are allowed, it is referred to as *net* royalty. The use of this method avoids the difficulties associated with discounting revenues and costs. However, the traditional royalty does have an impact on efficiency, for in the eyes of the producer, it adds to marginal costs. This occurs unless the royalty is calculated after allowances for *all* costs, including the supply price of capital. In all other cases, a wedge is driven between social and private costs of production, with the result that the amount of rent available for collection by a government is reduced. More importantly, perhaps, it must be noted that if a royalty is to be effective in collecting rent, the royalty rate must be calculated anew for each mineral deposit. Since rent is a function of differences in quality among mineral deposits, a standard royalty rate applicable throughout a jurisdiction will be unrelated to the rent derived from any single deposit.

The purpose of this discussion is not to compare the relative merits of different methods of collecting rent but to illustrate that the amount of rent available is not independent of the chosen method. It follows that, when the time does come for a comparison of methods, a judgment cannot be based simply upon the percentage of rent derived, even if it were possible to calculate this figure. One method of collecting rent may succeed in achieving a high percentage of what is available, but this sum may in fact be less (in absolute terms) than that obtained using a method collecting a lower percentage of a larger rent base.

This is the problem referred to by Paul Bradley in his article in this volume. When he talks about appropriating the rent of producing mines, he is concerned with the fact that the means of appropriation adopted in any time period influences the amount of the rent available in succeeding periods. Thus, he is prepared to abandon *short-term rent capture* in favour of *long-term revenue maximization* as the appropriate government objective in dealing with mineral resources. The same problem is also referred to by Ross Garnaut and Anthony Clunies Ross in their article in this volume, although the suggested solution is different.

David Quirin and Basil Kalymon state in their article in this volume that social welfare ought to replace rent collection as the appropriate government objective. This is equivalent to saying that a government should be guided by the principles of economic efficiency, so that its decisions ensure that the total social product is as large as possible. If it were not for one difficulty, this would be the same as saying that rent should be maximized and no venture should be undertaken which does not produce a net social benefit. The difficulty relates to the distribution of the total social product. If this could be shared amongst members of society without any loss of efficiency, Quirin and Kalymon's criterion would produce a result no different from that obtained where the objective was rent maximization. However, as already suggested, different methods of rent collection by a government have different implications for efficiency. None are entirely costless. A government is therefore in a position where it is forced to choose between efficiency and equity. Quirin and Kalymon argue that it is better for the government to emphasize efficiency at the expense of equity. This is where they part company with most of the other contributors.

The dilemma facing a government in control of publicly owned mineral resources is obvious. To the extent that there is any rent produced by development of these resources, the government has a strong claim in equity to it. However, it must be recognized that collection of this surplus involves a cost to society. This cost is likely to vary depending on the method of collection. The government must therefore strike a balance between rent collection and the size of the social product derived from mining and related activities. A number of factors will be important in achieving such a balance. For example, if there is a substantial degree of foreign ownership in mining, there will be a strong likelihood that most of the rent will go to people who are not members of the society which owns the resources in the ground. This will encourage that government to pursue rent collection avidly, even at the expense of economic efficiency. On the other hand, if the rent is likely to go to members of the society which owns the resources, the government will be strongly influenced by the distribution of benefits among such members of this society. If the likely distribution is in accord with government policy, the government may well choose to seek efficiency at the expense of government revenue. In any event, there is a strong argument for the type of research advocated by Paul Bradley in his article in this volume, since it is clear that the choice between available leasing mechanisms cannot be made on qualitative grounds alone.

Notes

1. Adam Smith, *The Wealth of Nations* (New York: Random House, 1937), pp. 144, 167-74.
2. T.R. Malthus, *Principles of Political Economy*, 2nd ed. (London: William Pickering, 1836), p. 136.
3. David Ricardo, *The Principles of Political Economy and Taxation* (London: Dent, 1969), pp. 46-47.
4. Ibid., p. 33.
5. John Stuart Mill, *Principles of Political Economy* (London: Longmans, Green & Company, 1909), p. 474.
6. Alfred Marshall, *Principles of Economics* (London: Macmillan, 1947), pp. 438-39.
7. Joan Robinson, *The Economics of Imperfect Competition* (London, Macmillan & Co., 1954), pp. 102. ff. This had been fully developed earlier by Lewis C. Gray, "Rent under the Assumption of Exhaustibility," *Quarterly Journal of Economics* (May 1974).
8. During the open session on the last day of the conference, Walter Mead suggested the contrary, *viz*, that a government should seek to capture the difference between *social* revenues and costs (see edited transcript pp. 259-61).

Biographical Notes

Paul G. Bradley is professor in the Department of Economics, University of British Columbia.

Michael Crommelin is a senior lecturer in law, University of Melbourne.

Harry F. Campbell is assistant professor in the Department of Economics, University of British Columbia.

Gregg K. Erickson is director of research, the Alaska State Legislature.

Mason Gaffney is professor in the Graduate School of Administration, University of California, Riverside.

Ross Garnaut is with the Research School of Pacific Studies, the Australian National University.

John Helliwell is professor in the Department of Economics, University of British Columbia.

Glenn P. Jenkins is an associate of the Harvard Institute for International Development.

Dale R. Jordan is a mineral management consultant with Seaton-Jordan & Associates, Calgary.

B.A. Kalymon is professor in the Faculty of Management Studies, University of Toronto.

Hayne E. Leland is associate professor in the School of Business Administration, University of California, Berkeley.

Brian W. Mackenzie is professor in the Department of Mining and Metallurgical Engineering, McGill University.

J.L. McPherson is vice-president, finance and administration, with Placer Development, Vancouver.

Walter J. Mead is professor in the Department of Economics, University of California, Santa Barbara.

A. Milton Moore is professor in the Department of Economics, University of British Columbia.

O.E. Owens is managing director of Cominco Europe in Vancouver.

Frederick M. Peterson is assistant professor in the Department of Economics, University of Maryland.

G.D. Quirin is professor in the Faculty of Management Studies, University of Toronto.

Anthony Clunies Ross is a lecturer in the Department of Economics, University of Strathclyde.

Anthony Scott is professor in the Department of Economics, University of British Columbia.

Andrew R. Thompson is professor in the Faculty of Law, University of British Columbia.

Arlon R. Tussing is a consulting economist to the United States Senate, Committee on Interior and Insular Affairs.

Russell S. Uhler is professor in the Department of Economics, University of British Columbia.

Leonard Waverman is professor with the Institute for the Quantitative Analysis of Social and Economic Policy, University of Toronto.

T.A. Wedge is a consulting mining economist in Toronto.

Index

Alaska, 12, 155, 166; North Slope, 36, 60, 190; Trans-Alaska pipeline, 169
Alberta, 10, 13, 92, 142, 143, 182, 185, 187, 194, 203, 204, 243, 244, 245, 246, 249, 250, 251, 254
ALCAN Highway, 165
Allais, Maurice, 35
Amortization, 17; allowance for, 8-10
Appalachia, 39
Arctic, 13, 35, 133, 142, 154, 156, 252
Area of mineral lease, 15, 27, 39, 168, 248, 249
Arrow, K.J., 137, 180
Atlantic Richfield Company, 36, 150
Averch-Johnson effect, 8
Azis, A., 218

Baumol, W.J., 137
Bentsen, Senator, 8
Bellman, R., 134
Bernoulli, Daniel, 153
Bonus bidding. See Cash bonus bidding
Britain, 61, 164, 165, 166, 178-79
British Columbia, 27, 35, 36, 37, 38, 39, 147, 149, 150, 151, 152, 163, 165, 166, 169, 171, 174-75, 177, 180, 185, 188, 190, 204, 206-8, 211, 213-14, 216-18, 220, 224, 258, 266-67, 268, 271; coal leases, 10; crude oil royalties, 13, 250, 253, 254; Department of Mines and Petroleum Resources, 149, 243; gas pricing, 18; Mineral Land Tax Act, 228; mineral leasing, 17, 227-30, 233-42; mineral royalties, 19, 38; Mineral Royalty Act (Bill 31), 19, 191, 228, 234; open access to mineral rights, 17, 227, 228, 229; Petroleum and Natural Gas Act, 248, 249, 250; petroleum leasing, 243-54; property tax, 12
British Columbia Hydro and Power Authority, 194
British Columbia Petroleum Corporation, 7, 13-14, 19, 175, 186, 187, 194, 252, 253
British Columbia-Yukon Chamber of Mines, 166
British North America Act, 21-22, 196, 198, 203

California, 151
Canada, 13, 40, 47, 95, 106, 113, 114, 133, 147, 156, 163, 166-67, 169, 170, 171, 180, 181, 182-98, 200-204, 205, 213-14, 218, 229, 233, 234, 235, 247, 251, 252, 260; Arctic oil exploration, 13; Bill C-259, 194; budget of 6 May 1974, 3, 11, 186-98, 242; Carter Commission, 202; Department of Energy, Mines and Resources, 194; federal income tax, 3, 4, 10-12, 14, 21-22, 114, 115, 116, 139-41, 169, 183, 184, 185, 186-98, 202, 203, 218, 228, 229, 233, 247; Fiscal Arrangements Act (1972), 185; James Bay, 133; Mackenzie Delta, 133; mineral leasing policy, 15, 18; mineral revenue policy, 194-95; National Oil Policy, 184, 192, 193-94; oil and gas pricing policies, 18. See also Alberta; British North America Act; British Columbia; Federal-provincial conflict over resource revenues; Manitoba; National Energy Board; Northwest Territories; Ontario; Saskatchewan; Yukon Territory
Canadian Bar Association, 230
Canadian Development Corporation, 196
Canadian National Railways, 164
Canadian Pacific Railway, 15, 183, 265, 266
Capital, 3, 6, 7, 8, 9, 10, 11, 12, 13, 14, 16, 17, 18, 20, 22, 23, 24, 25, 27, 79, 83, 84, 98, 99, 100, 101-2, 116, 133, 137, 140, 145, 155, 163, 164, 165, 167, 169, 170, 176, 178, 179, 183, 184, 188, 192, 197, 201, 202, 220, 230, 232, 234, 242, 257, 263, 269, 271, 273, 274, 275, 276, 277; cost of, 8, 9, 14, 16, 101, 103, 111, 115, 121, 123, 136, 168, 184, 188, 201; fixed, 6, 10; returns to, 20, 55, 176, 186-90, 257; recovery of, 9, 10, 20, 50; working, 6, 97. See also Investment
Cartel. See Market power
Cash bonus bidding, 22, 23, 31, 35, 36-37, 39-40, 46-60, 61-65, 89-90, 92, 99, 100, 104, 126, 143, 159-60, 168, 176, 179, 180, 190-91, 192, 231, 248, 249, 251, 254, 260, 262, 263, 277; advantages of, 50, 53-56; disadvantages of, 50-51; joint bidding in,